how
to
know
the
true bugs
(Hemiptera—Heteroptera)

The **Pictured Key Nature Series** has been published since 1944 by the Wm. C. Brown Company. The series was initiated in 1937 by the late Dr. H. E. Jaques, Professor Emeritus of Biology at Iowa Wesleyan University. Dr. Jaques' dedication to the interest of nature lovers in every walk of life has resulted in the prominent place this series fills for all who wonder **"How to Know"**.

John F. Bamrick and Edward T. Cawley
Consulting Editors
Louise K. Barrett
Editor

The Pictured Key Nature Series

How to Know the

- **AQUATIC PLANTS,** Prescott
- **BEETLES,** Jaques
- **BUTTERFLIES,** Ehrlich
- **CACTI,** Dawson
- **EASTERN LAND SNAILS,** Burch
- **ECONOMIC PLANTS,** Jaques, Second Edition
- **FALL FLOWERS,** Cuthbert
- **FRESHWATER ALGAE,** Prescott, Second Edition
- **FRESHWATER FISHES,** Eddy, Second Edition
- **GRASSES, Pohl,** Second Edition
- **GRASSHOPPERS,** Helfer, Second Edition
- **IMMATURE INSECTS,** Chu
- **INSECTS,** Jaques, Second Edition
- **LAND BIRDS,** Jaques
- **LICHENS,** Hale
- **LIVING THINGS,** Jaques, Second Edition
- **MAMMALS,** Booth, Third Edition
- **MARINE ISOPOD CRUSTACEANS,** Schultz
- **MOSSES AND LIVERWORTS,** Conard, Second Edition
- **NON-GILLED FLESHY FUNGI,** Smith-Smith
- **PLANT FAMILIES,** Jaques
- **POLLEN AND SPORES,** Kapp
- **PROTOZOA,** Jahn-Jahn, Second Edition
- **ROCKS AND MINERALS,** Helfer
- **SEAWEEDS, Abbott,** Second Edition
- **SPIDERS,** Kaston, Second Edition
- **SPRING FLOWERS,** Cuthbert, Second Edition
- **TAPEWORMS,** Schmidt
- **TREMATODES,** Schell
- **TREES, Miller-Jaques,** Second Edition
- **TRUE BUGS,** Slater-Baranowski
- **WATER BIRDS,** Jaques-Ollivier
- **WEEDS, Wilkinson-Jaques,** Second Edition
- **WESTERN, TREES,** Baerg, Second Edition

how
to
know
the
true bugs
(Hemiptera—Heteroptera)

James A. Slater
University of Connecticut

Richard M. Baranowski
University of Florida

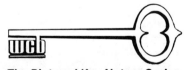

The Pictured Key Nature Series
Wm. C. Brown Company Publishers
Dubuque, Iowa

The following figures are reproduced by permission of the institution or individual as indicated below.

Figures: 420—University of California Press.
457, 458, 464, 465, 471 through 473, 475 through 480, 482 through 484—University of Kansas.
26, 32 through 35, 57, 64, 80, 89, 91, 96, 97, 102, 120, 136, 139, 194, 195, 202, 248, 264, 268, 419, 426, 439, 447 through 455, 474, 485, 487, 492—Dr. R. C. Froeschner.
421 through 425—American Entomological Society.
281, 303, 314, 316, 335, 340, 341, 368, 378, 385, 387, 388, 402, 403, 411, 412—Dr. J. C. M. Carvalho.
277, 280, 293, 297, 298, 302, 307, 310, 313, 315, 320, 322, 331, 339, 342, 360, 369, 376 379, 380, 382, 383, 384, 389, 391, 396, 399, 404, 407, 410, 415—Illinois Natural History Survey.
103—New York Entomological Society.
428 through 431—Carnegie Museum of Natural History.
196, 198, 200—Iowa State University.
214, 215—Philippine National Institute of Science and Technology.
337, 346 through 350, 390—Entomologist's Monthly Magazine.
12 through 15, 17 through 22—Florida Entomologist.
381, 392, 397, 409—Dr. Maldonado-Capriles.
259 through 263, 265, 266—Entomological Society of America & Thomas Say Foundation.
244, 245, 247, 249—American Museum of Natural History.
3 through 6, 10, 11—D. F. Waterhouse and Melbourne University Press.
1, 2—Connecticut Geological and Natural History Survey.
23, 187, 189—British Museum (Natural History).
469—Mr. Steve Gittelman.
278, 279, 358, 377, 393, 395, 400, 401, 406—Brigham Young University.

Copyright © 1978 by Wm. C. Brown Company Publishers

Library of Congress Catalog Card Number: 76-24534

ISBN 0-697-04893-4 (Cloth)
ISBN 0-697-04894-2 (Paper)

All rights reserved. No part of this publication may be reproduced, stored in a retrieval system, or transmitted in any form or by any means, electronic, mechanical, photocopying, recording, or otherwise, without the prior written permission of the publisher.

Printed in the United States of America

Contents

Preface vii
Acknowledgments ix
The Order Hemiptera 1
   The True Bugs 1
   Life Cycles of Hemiptera 2
   Morphology 3
Collecting 7
   Equipment 7
   Terrestrial 9
   Aquatic 10
Preservation and Labeling 12
The Workers 16
   The Europeans 16
   The Classic Americans 17
General References 19
Key to Adults of North American Families of Hemiptera 20
Key to Nymphs of North American Families of Hemiptera 26
Families of Hemiptera 33
   Cydnidae 33
   Thyreocoridae 37
   Pentatomidae 40
   Coreidae 57
   Alydidae 63
   Rhopalidae 66
   Lygaeidae 70
   Pyrrhocoridae 95
   Largidae 96
   Piesmatidae 97
   Thaumastocoridae 98
   Berytidae 99

Aradidae 103
Tingidae 106
Enicocephalidae 117
Phymatidae 118
Reduviidae 119
Nabidae 134
Polyctenidae 139
Cimicidae 139
Anthocoridae 143
Microphysidae 147
Miridae 147
Isometopidae 205
Dipsocoridae 206
Schizopteridae 207
Hydrometridae 209
Gerridae 210
Veliidae 214
Mesoveliidae 217
Hebridae 218
Leptopodidae 219
Saldidae 220
Notonectidae 224
Pleidae 230
Corixidae 230
Nepidae 238
Belostomatidae 239
Naucoridae 242
Gelastocoridae 243
Ochteridae 244

List of Common Names 245

Index and Glossary 249

Preface

The need for a synoptic manual to aid both beginning students and professional entomologists in the identification of the true bugs of the order Hemiptera has been apparent for many years. The existing literature is very widely scattered in both American and European scientific publications, and many of our most common and widespread species have never been adequately illustrated. The last comprehensive work for the eastern United States was published in 1926, and for the western states only the aquatic and semi-aquatic bugs have ever been treated in a single volume.

One of the most difficult problems we have had to face in preparing this manual has been which species and which genera to include and which to exclude. Ideally a manual of the true bugs should include all of the known species. However, such a volume would of necessity be many times the size of this book. We have, therefore, had to compromise throughout. In general we have tried to include the most common and widespread species. However, we are the first to admit to some inconsistency in this, for we have also tried to include many of the bizarre, striking or biologically unusual species even though many of them may be rare or of restricted distribution. The student using this manual should be aware that not all species and genera are included, and he will, no doubt, sometimes be frustrated to find that he has a genus or species before him that is not treated in the book. However, it is our hope that this manual will aid the beginning student and general entomologist to become familiar with this important and fascinating group of insects, and that it will serve as a stimulus for students at all levels to proceed from it to the original and definitive literature.

We have given the general distribution of all the species included, but it must be admitted that our knowledge of the distribution of most true bugs remains fragmentary. The given ranges should not be considered accurate in the same sense as are those of the better known vertebrate animals, but rather should be used as general approximations of distribution. Any serious student will find many species existing outside of the ranges listed here.

We have also attempted to indicate the approximate number of additional species that are known in a given genus other than those treated in the present manual.

In treating the great family Miridae, which contains by far more species than any other family, we have deviated from the usual format. In this family we have attempted to produce a key which includes all of the North American genera, but we have included only the most common and distinctive species. This family is so extensive, and identification of species so difficult, that we have felt it more useful to enable a student to identify specimens to genus only and thus encourage him or her to turn to the primary literature for species determination. Even at the generic level the key probably will not always prove to be satisfactory as there are a number of genera and many species that we have not seen, and frequently we have had to work from published information only.

Since higher categories of classification are subjective there are several current disputes in Hemipterology over the status of "family" units. These involve, for example, whether the Acanthosominae and Scutellerinae should be considered to be subfamilies of the inclusive family Pentatomidae or whether both or either of them deserve family status. The same holds true of the Phymatidae relative to the Reduviidae, the Macroveliidae to the Veliidae, the Anthocoridae to the Cimicidae, the Berytidae to the Lygaeidae and the Pleidae to the Notonectidae. We have probably not always been consistent in treatment in the present work, but our intent has been to adopt a conservative approach. Manuals of the type offered here are, in our opinion, not the place to introduce or quickly adopt radical changes in the family status of various well-known groups. This tends to confuse the beginning student and makes it difficult for him or her to use the past literature. At the same time it does not add materially to the knowledge of the advanced worker who is working from the primary literature and not from manuals. Therefore, in most cases we have retained the family units much as they appear in most standard American works.

Acknowledgments

Many persons have aided us in the preparation of this work, and to them and to the work of the Hemipterologists of the past goes the credit for whatever utility this manual may have. To the following persons we are deeply indebted for supplying us with specimens, testing the keys, suggesting changes and what material to include and exclude: P. D. Ashlock (University of Kansas), N. P. Chopra (Haryana Agricultural University), R. C. Froeschner and J. L. Herring (National Museum of Natural History), R. I. Sailer (University of Florida), J. D. Lattin (Oregon State University), Veronica Picchi and Diane Calabrese (University of Connecticut), J. Polhemus (Englewood, Colorado), J. C. Schaffner and M. H. Sweet (Texas A. & M.), P. Wygodzinsky and R. T. Schuh (American Museum of Natural History) and H. V. Weems and F. W. Mead (Florida Department of Agriculture and Consumer Service, Division Plant Industry).

We hope that much of the value of this book will lie in the illustrations, and we wish to express our deepest gratitude to the artists Mrs. Sally Kaicher, Dr. Jack Bacheler (North Carolina State University), Mrs. Karen Velmure (Gray Herbarium), and Mrs. Kathleen Schmidt (University of Connecticut) for their patient attention to our many needs in the production of the text illustrations.

Our appreciation is extended to the following for permission to use figures from their works:

The University of California Press for the use of "The Aquatic Insects of California" by R. L. Usinger; the University of Kansas for the use of "The Corixidae of the Western Hemisphere" and "The genus Notonecta of the World" by H. B. Hungerford; Dr. R. C. Froeschner for the use of original drawings from "The Hemiptera of Missouri" and "Cynidae of the Western Hemisphere" and for the use of the original drawings of Saldidae executed by Mr. Arthur Smith for the late Professor C. J. Drake; the American Entomological Society for the use of "The Schizopteridae, with the Description of New Species from Trinidad" by M. G. Emsley; J. C. M. Carvalho for the use of "Keys to the Genera of Miridae of the World"; the Illinois Natural History Survey for the use of "The Miridae of Illinois" by H. H. Knight; New York Entomological Society for the use of an article from the Journal by R. F. Hussey; the Carnegie Museum of Natural History for the use of an article from the Annals by C. J. Drake and H. M. Harris; the Iowa State University for the use of an article from the Iowa State College Journal of Science by M. P. Hurd; the Philippine National Institute of Science and Technology for the use of "Revision of Phymatidae (Hemiptera: Phymatidae)" by N. A. Kormilev; the Entomologist's Monthly Magazine for the use of an article by J. C. M. Carvalho and D. Leston; the Florida Entomologist for the use of an article by J. Herring and P. D. Ashlock; Dr. Maldonado-Capriles for the use of original figures; the Thomas Say Foundation and the Entomological Society of America for the use of "Monograph of Cimicidae" by R. L. Usinger; the American Museum of Natural History by financial agreement for the use of "Monograph of thte Emesinae" by P. Wygodzinsky; D. F. Waterhouse and Melbourne University Press for the use of "The Insects of Australia"; the Connecticut Geological and Natural History Survey for the use of "Hemiptera of Connecticut"; the British Museum (Natural History) by financial agreement for the use of "Classification of the Aradidae" by R. L. Usinger and R. Matsuda and Brigham Young University for the use of illustrations from "Miridae of Nevada Test Site and the Western United States" by H. H. Knight.

Our sincere thanks are also due to the late Mrs. Darleen Wilcox (University of Connecticut) and to Mrs. Elizabeth Slater and Miss Jane O'Donnell (University of Connecticut) for extensive aid in testing the keys and for editing of the manuscript.

Finally we express our appreciation to the administrations of the University of Florida and the University of Connecticut for financial aid during the course of the work.

The Order Hemiptera

THE TRUE BUGS

The study of any group of insects is a fascinating and challenging experience and is a source of pleasure that constantly increases as one becomes more and more familiar with the various species and their great variety of habits and adaptations to a bewildering number of living places. Although the size, beauty, and diversity of butterflies, moths, beetles, and dragonflies often first attract the young naturalist, many of the other orders offer equally interesting fields of study. High among these are the true bugs of the order Hemiptera which are treated in this book. We have studied these insects in field and laboratory for many years and remain constantly astonished by their diversity, beauty and by the almost infinitely large number of problems remaining to be solved in the biology and classification of even the most common species. It is our hope that this volume will enable the beginning student to identify these insects, to enjoy them as we have enjoyed them and lead him, whether as a vocation or avocation, to an exploration of their biology and ecology.

It may truly be said that there is scarcely a single habitat except the depths of the sea itself that does not have an hemipteran exquisitely adapted to live there. No other insect order is represented in such a variety of habitats as are the hemipterans. Probably the majority of species are plant feeders sucking the juices from leaves, stems, berries, roots, flowers, and seeds. Among these, certain infamous species such as the tarnished plant bug, chinch bug, harlequin bug, and squash bug stand out because of the considerable damage they inflict on economic crops. Thousands of species are predatory on other insects. Notable among these are *Podisus placidus* which feeds on fall webworms and tent caterpillars and *Perillus bioculatus* which feeds on the larvae of the Colorado potato beetle. In addition many other stink bugs as well as assassin bugs, damsel bugs, and minute pirate bugs (flower bugs) are also predatory on other insects. Some hemipterans, such as the bedbugs and batbugs, are parasites upon man, birds, and bats.

Despite their abundance and diversity, surprisingly little is known about them and one can hardly spend a summer in the field anywhere in North America without being able to add many new facts to our knowledge of these insects. We hope users of this book will not be content merely to learn the names of the true bugs they collect but will use this knowledge to obtain the badly needed information on life histories, ecology, and distribution that provide so much intellectual pleasure as well as contributing to the growth of science.

What is a true bug? This might seem to be a simple thing to define, yet it is quite difficult. The various species have been modified so greatly in response to environmental specialization that one can hardly find a single characteristic that every true bug has. Most text books will tell you that the Hemiptera may be recognized by having a distinct, segmented beak arising from the front of the head and extending backward along the underside of the head; by having the front wings lying nearly flat over the abdomen with the outer half of the wings overlapping one another and being much thinner than the basal half. Indeed, it is this latter feature that gives the order its name "Hemiptera" or "Half-wings" which, of course, doesn't mean that only half a wing is present but rather that the texture of the wing is of two dissimilar halves. It is true that these two characteristics will enable one to identify the great majority of the true bugs that are encountered. Yet, many adult hemipterans are entirely wingless, others have the wings reduced to tiny flap-like pads that scarcely extend onto the abdomen, others have the forewings shell-like and resemble small beetles, still others have lacy wings with numerous cells of the same texture throughout, while in others the usually triangular scutellum becomes large and hemispherical and completely covers the wings. Still, if you learn the beak and wing characteristics you will be able to recognize 90% of the true bugs as such at a glance and the beak alone will enable you to tell almost all except the highly specialized water boatmen whose mouthparts look like nothing more than a few bristles.

The order name itself causes many students perplexity and comes about because some entomologists think the true bugs and relatives such as the cicadas, leafhoppers, aphids, scale insects, etc., all belong to one order which they call the Hemiptera, with two suborders the Homoptera for the cicadas, aphids, etc. and the Heteroptera for the true bugs. Other entomologists feel that each of these groups represents a distinct order and then the true bugs are called the Hemiptera and the cicadas, aphids and relatives Homoptera. Thus Hemiptera can be used in two connotations. This really makes little practical difference as anyone can quickly tell which concept is involved in a given work, but it does cause problems to beginning workers trying to go to different books (and to computers). It does have the practical value of allowing entomologists to argue about the relative objectivity of subjective categories.

LIFE CYCLES OF HEMIPTERA

All members of the order have an incomplete type of metamorphosis in that there is an egg stage, a series of nymph (=larva) or growing stages and an adult stage.

Eggs of Hemiptera may be inserted in plant tissues, placed in masses on stems and leaves, inserted in crevices in bark or on the ground, or merely dropped at random on the substrate. While many eggs are of simple elliptical shape some are quite bizarre, those of Pentatomidae resembling small barrels and those of Reduviidae ornamented on the opercular surface with whorls of spines and filaments.

Nymphs almost invariably pass through five instars (the insect between two molts) during which time the size increases greatly. Wing pads usually are visible as small flaps on the caudo-lateral angles of the mesonotum by the third instar and become progressively larger in the fourth and fifth instars. A few cases are reported where particular species pass through only four or conversely as many as six instars, but the consistency of five instars is remarkable for such a large and diverse order.

Nymphs of the majority of species have scent glands on the dorsal surface of the abdomen. In adults these abdominal glands are absent but their position may often be ascertained by "scars." The nymphal scent glands are presumably for protection.

Adults of all or nearly all Hemiptera feed actively and live for a considerable period of time. Males and females may be distinguished externally by examining the end of the abdomen. The male abdomen generally terminates in a round or irregular capsule whereas in the females a series of flat plates or elongate blades are usually present.

MORPHOLOGY

The study of the shape and function of the parts of insects is in itself an important division of entomology. Indeed, many scientists devote a major portion of their careers to morphological studies seeking to unravel the evolutionary relationships between groups of insects.

However, to identify insects it is only necessary to understand and recognize relatively few features including some that are special to the bugs.

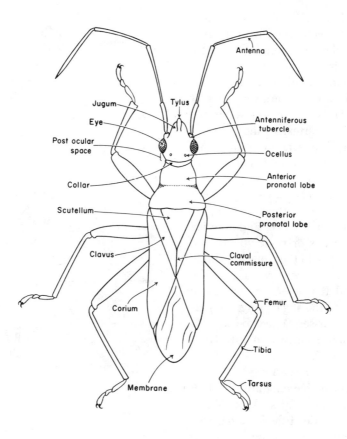

Figure 1

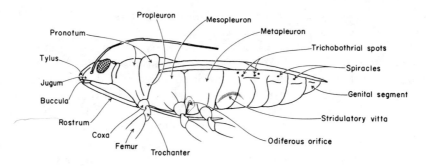

Figure 2

As with other insects the body is divided into three regions, the head, thorax and abdomen.

Head (Figs. 1, 2, 3, 4).—The shape and size of the head varies considerably in the true bugs, but in nearly all cases the most prominent structures are the pair of large compound eyes, each occupying a large portion of the side of the head and made up of a large number of facets or ommatidia. The area between the compound eyes is the vertex. In most bugs a pair of simple bead-like eyes or ocelli (Fig. 1) are located here. A pair of jointed antennae arise from the head in front of the eyes in most bugs. Usually the antennae are 4-segmented but are 5-segmented in some species. The portion of the head between the eyes and anterior to the vertex is the frons or postclypeus. The part of the head anterior to the frons has three, usually distinct, areas, the center is termed the tylus (clypeus) and the outer two areas are known collectively as the juga (singular jugum) (Fig. 1). The four-segmented beak (also called rostrum) or labium arises from the front part of the head (Figs. 2, 4).

Mouthparts (Fig. 5):—Almost all Hemiptera (see Corixidae) have mouthparts that are adapted for piercing and sucking. When one looks on the underside of the head one sees a "beak-like" structure. This is the labium. It has a groove along the (visually) ventral (morphologically dorsal) surface in which lie four

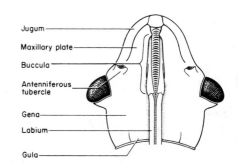

Figure 3

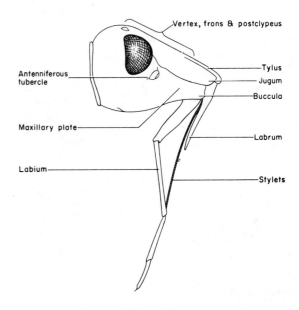

Figure 4

slender threads, or stylets; the outer pair are the mandibles, the inner pair the maxillae. Near the base of the labium a "flap" is present, the labrum. Each mandible has sharp teeth near the tip which cut a minute hole in the food source. The maxillae are longitudinally grooved on their inner surfaces and lie closely appressed so that the inner surfaces form two canals, the food canal and the salivary canal (Fig. 5bb[1]). In feeding, salivary fluid is pumped down the salivary duct and liquified food material is pumped up the food canal. The pumping action is accomplished by a complex arrangement of plates attached to the stylets that are continuous with a sucking pump in the head called a cibarium.

Thorax.—This region of the body is divided into three segments, the prothorax, mesothorax and metathorax, each of which bears a pair of legs. The front pair of wings, the hemelytra, are carried by the mesothorax and second pair of wings by the metathorax.

The top of the prothorax is called the pronotum (Fig. 2) and is quite variable in the various groups of bugs. The central part of the pronotum is often referred to as the disk; the outer edges as the lateral margins. The angles where the lateral margins meet the hind margin are the humeral or lateral angles, sometimes referred to simply as humeri. In some bugs there are two slightly raised areas near the front margin of the disk, these are the calli. The front portion of the disk is sometimes separated from the remainder by a depression or groove which is then referred to as the collar.

The wing-bearing segments, the mesothorax carrying the hemelytra or front pair of wings and the metathorax carrying the hind pair of wings are rather rigidly united since they contain the important flight muscles. Lying above the meso- and metathorax and posterior to the pronotum is a triangular structure, the scutellum.

The lateral portion of (Fig. 6) each thoracic segment is called a pleuron and ventrally a sternum. Usually the latero-ventral surface is divisible into two recognizable anterior and posterior areas, the episternum and the epimeron. Thus (Fig. 6) one can recognize in Hemiptera a proepisternum, proepimeron, mesepisternum, mesepimeron, etc. Between the middle and hind legs on the metapleuron

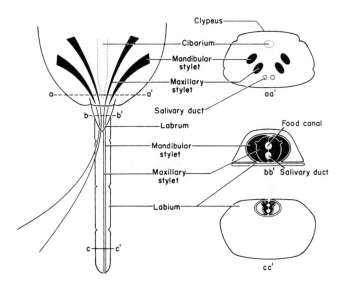

Figure 5

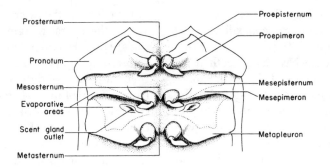

Figure 6

is an opening (often ear-like) which is the orifice of the adult scent glands. Around the orifice is a granular appearing region which is called the evaporative area.

The legs of bugs may be greatly modified for predation or swimming but typically the following parts can be recognized; the coxa or basal segment of the leg, the trochanter, a small sclerite occasionally fused to the femur which is a long segment usually heavier than the tibia. The last section is the tarsus, made up of 1-3 movable joints, which terminates with a pair of claws that may or may not have pads called arolia or parempodia, at their bases.

Abdomen.—This section of the body is made up of 9-10 segments each having a recognizable upper plate or tergum and a ventral under plate (sternum). The side margins of the abdomen at the junction of the dorsal and ventral plates are often flattened and prominent to form the connexivum. The 8th and 9th segments are modified to form the genitalia and have characters much used in classification.

Collecting

EQUIPMENT

In general the equipment used and techniques followed in collecting Hemiptera are similar to those utilized for many other insect orders. We suggest that the beginning collector consult a good introductory textbook for general details. (Borror, Delong and Triplehorn, "*An Introduction to the Study of Insects*", Rinehart and Co., N. Y. has an excellent introductory discussion). The following discussion is limited to details that we have found to be of particular value in Hemiptera collecting.

1. **Nets:** Most experienced collectors have particular types of nets that they favor. Our own experience has been that for general purpose work a rather heavy net with a short aluminum or wooden handle, heavy gauge wire loop, and heavy opaque bag with an elongate taper is the most effective. It should have a diameter of about 15″ with a handle approximately three feet long. The rim should be made of a heavy wire about 1/4″ in diameter. Preferably the bag should be about twice as long as the diameter of the rim and made of heavy unbleached muslin, drill, or light canvas. The handle can be obtained from a discarded broom or mop, the rim formed from number six or eight gauge wire. Although the muslin net can be used for aquatic collecting, a bag constructed of nylon having a mesh of 24 to 32 threads to the inch is better. The denser muslin causes a greater drag in the water making it more difficult to capture quick-moving insects.

2. **Aspirators** ("pooters"): This device which enables one to quickly capture small insects without touching them is a very valuable tool. Many scarce and interesting Hemiptera must be laboriously searched for on hands and knees by scratching under shrubs, grass clumps, and among tree litter. In areas where spines and thorns are abundant, or where the possibility of encountering scorpions is high, we find a pair of heavy "work gloves" to be comforting both physically and psychologically. For collecting of this type an aspirator is essential. In fact you will soon find that this device becomes a third hand and that without it you will feel virtually helpless in the field. In our experience the most effective aspirator (or "pooter") is made by utilizing two pieces of brass or copper tubing, one about eight inches long with an inside diameter of approximately 1/4″ and one slightly larger and about five inches long. These tubes are inserted in a two-hole rubber stopper, a fine wire mesh (similar to the

type used in gasoline funnels) is soldered over the end of the thicker tube and a rubber or plastic tube about three feet long is attached to the other end of the thicker tube. (A piece of nylon stocking can be used temporarily for mesh.) We have found that a clear plastic vial (15 dram vials obtainable at most drugstores) with a diameter of 1 1/2″ and length of 3 1/4″ that has a fitted plastic snap cap is very useful. These can be interchanged on the rubber stopper. The vials with insects may be stored safely and the insects either killed immediately or retained alive until returning to the laboratory.

3. **Beating tray:** This device consists of a piece of green or white cloth about a yard square with light wooden or metal cross rods fitted into "pockets" below each corner. This tray can be placed under bushes or other plants while they are shaken or struck with a stick. This is a particularly valuable tool in areas where thorny vegetation makes extensive use of the net impractical.

4. **Sifters:** Nearly any kind of container with a wire-mesh bottom can be used to sift small insects from debris. Small household strainers are especially handy for collecting aquatic insects.

5. **Separators:** Various types of separators have been devised to force insects out of leaf-mold or litter of various sorts. A Berlese funnel is the most common.

6. **Light traps:** Almost any source of light can be used to attract Hemiptera. An ultraviolet lamp, commonly referred to as a blacklight, is especially attractive to many insects. When electricity is not available a gasoline or battery powered lantern placed near a white sheet hung between two trees will provide many hours of profitable night collecting.

7. **Killing bottles:** Types of killing bottles vary with the collector. The traditional killing bottle mentioned by many books contains potassium-, sodium-, or calcium cyanide. This is effective and very long lasting. However, the danger to human beings is so great that we do not recommend its use. We find that ethyl acetate placed on cotton, cellucotton, or Plaster of Paris makes a safe and effective killing jar. Although bottles of nearly any shape and size can be used, straight-sided bottles about one to two inches in diameter and four to six inches long are most convenient. Corked bottles are also easier to use than screw-capped bottles.

A collector will also find it convenient to carry a pair of forceps to handle specimens, a hand lens, several spare vials, and a notebook or tape recorder in which to record data. All too often a person thinks he can remember the information until he gets out of the field only to realize later that he cannot. It is far easier and more accurate to record all data in the field. A small loose-leaf notebook is best for short trips. Recorded data from previous collecting trips should not be taken into the field. The possibility of losing valuable data is too high. Many collectors find it handy to carry their field equipment in a small army-type knapsack, gas mask, or musette bag. A bag that contains a strap that can be fastened around the waist as well as one that goes over the shoulder is best as it prevents the bag from swinging. These can usually be purchased at surplus stores. Plastic bags for litter samples and various containers for live material are also desirable field items.

TERRESTRIAL

A person becomes a successful collector only through the use of intelligence, hard persistent work, and keen observation. Although Hemiptera can be found in nearly every type of habitat one can spend many fruitless hours in the field attempting to collect in unfavorable locations. Thus a person should train himself or herself to recognize the collecting potential of an area. Skilled collectors often can tell how productive an area is likely to be after only a very few minutes observation. Walking through a field or along a roadside swinging a net is a very effective way to collect many insects but it should be remembered that by and large these will be the most common species.

Sweeping with a net is an effective method for collecting many Hemiptera from fields, shrubs, and trees. It is best accomplished by advancing slowly and swinging the net from side to side in a figure eight pattern. After several sweeps are taken in a desired location, the net should be swung in a manner to cause the lower half of the bag to loop over the rim to prevent captured insects from escaping. The net should then be slowly and carefully turned inside out. Insects trapped in the end of the net can then be removed by hand, by a forceps, or (we prefer this for most) if quite small, with an aspirator. Be particularly careful to examine the debris at the end of the bag as some relatively rare, minute species can readily be overlooked. Some collectors dump the entire contents of the net into a wide mouthed killing jar. Unfortunately this may result in a soggy mass of weed seeds, leaves, grass, and smashed specimens, particularly if not sorted within a short time.

The use of a net is not limited to sweeping fields. It can also be used to "beat" bushes and low branches of trees. One technique often used is to jar tree branches with a heavy stick while holding the net beneath the branch to catch the falling insects. This is also an effective method of capturing insects that secrete themselves in the inflorescences of tall grasses and sedges.

The most productive collecting will be realized by paying particular attention to the various habitats in an area. These may range from microhabitats, such as in flowers or the bases of various species of sedges or grasses, to larger habitats, such as brushy roadsides and weedy fields. By slowly searching the various habitats the collector is able to find the less common species and in addition is able to obtain valuable ecological and host plant data. The host plants and immature stages of a great many Hemiptera are unknown and special care should be given to establishing host plants and associating nymphs and adults.

While it is not possible to give collecting hints that will cover all possibilities, some are given under the discussion of individual families and a few others are given here. A careful search of flower heads will often reveal phymatids waiting to prey on other insects. Often tiny anthocorids and many mirid species are found in flowers. Examining the bark of trees and pulling loose bark off dead and decaying trees will produce anthocorids, pentatomids, aradids, reduviids and, occasionally very rare isometopids. The undersurface of leaves of various trees and shrubs often harbor tingids. Since most of these species are also host specific, knowing the plant on which the specimens are found will help in the identification. Other Hemiptera that may be found feeding on the foliage of shrubs and trees are pyrrhocorids, pentatomids, largids, coreids, alydids, berytids, and mirids. Predatory insects such as the reduviids, nabids, and some pentatomids will also be found on foliage, frequently where large numbers of other insects occur.

One of the most rewarding methods of collecting might be termed the "scratch and search" technique. The collector finds himself in the rather undignified position of being on his hands and knees with his face quite close to the ground, an aspirator in one hand and scratching with the other. (One becomes accustomed to the stares and questions of passersby very rapidly!) One says "I've lost a ring" or "I'm trying to reunite Gondwanaland," etc. Decaying vegetation, fallen seeds, fruits, and various small organisms that are associated with ground litter provide an excellent microhabitat for many species of Hemiptera, and many unusual species confine themselves almost exclusively to such habitats and to the bases of clumps of grasses and sedges. The usual procedure is to scratch gently through the debris found on the ground and aspirate the specimens as they run for cover. Rapidly agitating the debris in an area of about one square foot and then waiting a few moments will often produce surprising results as the "shaken up" specimens run for cover. This type of collecting can be physically hard since you may be in a cramped position for long periods and is often not used sufficiently because the collector doesn't "cover much ground" but from the standpoint of the species captured it is highly rewarding.

Since very few collectors can identify all of the plants that they may come in contact with on a collecting trip, it is often necessary to collect plant specimens for future identification. In the case of small plants, the entire plant may be taken, but for larger ones representative samples of the leaves, twigs, and flowers should be obtained. It is always desirable to obtain flowers and roots if available, particularly of grasses, for often this is the only way of identifying the species. If a collector expects to be away from home base for a few days a plant press should be taken into the field. If you are going to be in the field for only a few hours the plants can be kept in plastic bags.

It cannot be emphasized too strongly that the specimens collected are only as good as the data collected with them. A specimen with a label indicating that it was collected on goldenrod is much more valuable than one without any host data. Even a label "sweeping weeds" is better than none at all. But it is especially important with phytophagus species to indicate if nymphs were present as this establishes a true "breeding" population.

AQUATIC

Since many species are very restricted in where they are found it is equally important to pay close attention to the habitat when collecting aquatic Hemiptera. One can obtain many species by sweeping through the water with an aquatic net and pushing the rim of the net against underwater vegetation to dislodge resting insects. Forms which inhabit streams may be captured by holding the net vertically in the water and dislodging rocks and jarring vegetation just upstream of the net. Those species that inhabit the shoreline and water's edge have to be searched for in the moss, among rocks, and among the marginal plants. Some collectors splash water up on the shore to force species out of the vegetation. A small tea strainer is useful in this type of collecting. To capture swift moving water striders and jumping forms requires a great deal of skill and ingenuity. Collectors soon work out their own special methods of capturing these insects.

There are many types of specialized equipment such as Needham scrapers, various dredges, and samplers used for collecting aquatic insects available in most laboratories. A good discussion of their use can be found in Usinger (1963).

Many species in the genera *Hebrus, Merragata, Microvelia, and Paravelia* as well as mesoveliids, gerrids, and hydrometrids are found on the water surface and on and around emergent vegetation. Members of the genus *Rhagovelia* are also found on the surface, but they inhabit the riffles in streams often darting about in schools.

Toad bugs of the genus *Nerthra* are very secretive, living under debris, logs, and stones usually in damp situations, but occasionally a considerable distance from water. Those in the genus *Gelastocoris* and related ochterids are usually found on sandy or gravelly shores of streams or lakes. Pleids and notonectids are found primarily in lakes and ponds usually associated with submerged vegetation and are best collected by sweeping. Belostomatids, naucorids, and corixids are also found under these conditions. Water scorpions are poor swimmers and are usually clinging to submerged vegetation. Vigorous sweeping is necessary to dislodge them.

Saldidae often occur on open mud flats, beaches, and rocks in streams. They are extremely agile and fly readily. Dr. J. Lattin and Dr. T. Schuh have written that they have been able to capture most species readily as follows: "a 250 ml. laboratory squeeze bottle filled with 70% alcohol and an aspirator with a long reach are indispensable to the saldid collector. The rapid flits of the shorebugs usually leave the collector with little more than an aspirator full of sand unless a 'squirt gun' is employed. By first locating a saldid on the substrate and carefully advancing to within about a foot, a quick squirt of alcohol from the squeeze bottle can be used to stop the saldid short so it can be easily picked up with the aspirator. After being hit with the alcohol the bug will often jump around erratically for a moment; therefore it is necessary to keep your eyes carefully fixed on it."

Many Hemiptera may be placed directly into 70% ethyl alcohol until it is convenient to mount them. All nymphs should be preserved permanently in such a solution. However, preservation in alcohol is undesirable for certain species. Most Miridae are extremely delicate and alcohol preservation should be avoided. Species with bright coloration particularly reds, greens, and orange often fade badly in alcohol. Frequently alcohol preservation causes the wings, of pentatomids and coreids particularly, to partially open which some collectors consider undesirable. Only experience will enable the collector to know where alcohol preservation is desirable and where it should be avoided. If specimens that cannot be placed in alcohol cannot be mounted immediately they should be placed in pill boxes on cellucotton or soft tissues (cotton should not be used as the claws cling to it and result in the frequent loss of legs). Whenever possible Miridae should be mounted fresh. Not only does this reduce the danger of appendage breakage of these extremely delicate insects but the pubescence on the dorsal surface is very easily abraded and is important in generic and specific identification.

Preservation and Labeling

It is an entomological cliché that an hour of collecting requires at least two hours of preparation of specimens. The beginning student should always keep in mind that the mere accumulation of a large number of species and specimens is in itself of limited value. The truly good collection is one where every specimen is neatly and accurately mounted with detailed data attached so that both the collector himself and other investigators are able to obtain a maximum amount of information as quickly and accurately as possible.

Mounting: (Fig. 7) Most adult hemipterans are either glued to points or, if sufficiently large, a pin is run directly through the body. The question is often asked whether medium-sized insects should be pinned or point mounted. The best rule of thumb is "if doubt arises always point mount." (a) Point mounting: The insect is glued to the side of a simple triangular "point." This requires a small amount of practice. The best paper we have found for "points" is Strathmore board which can be obtained at most bookstores and artist supply houses. Points should be from seven to ten mm long. The best points are those that are punched out to uniform size by a commercially made punch. Commercial punches are available that produce either a blunt tipped point for larger insects or a sharp tipped point for very small insects. While a punch costs a few dollars it is well worth the investment because of the uniformity and neatness that results. Triangles or "points" can of course be cut individually by hand with scissors. The tip of the point should always be bent down at a right angle or "crimped." Ordinary glue may be used to fasten the insect to the "crimped" end of the point (some swear by "Elmer's Glue-all") but many types of glue tend to become brittle with

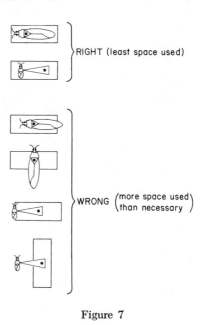

Figure 7

age. Clear fingernail polish (amyl acetate) eliminates this problem and is an excellent substance for point mounting. The only disadvantages of fingernail polish are that it sets more slowly than some glues and is not water soluble. A material called "Gelva" sets more quickly and firmly than fingernail polish and does not appear to become brittle with age. The point should be oriented on the pin so that it projects to the left when viewed from above. The adhesive substance is placed on the "crimped" end and this pressed firmly against the *right* side of the insect, usually against the thorax. The specimen should then be carefully teased into a position where the dorsal surface is level when viewed from above, the insect is thus on the left side of the pin with the head projecting away from the observer. The little extra time spent in accurately orienting the specimen is well repaid by the resulting appearance of the collection. (b) Body pinning: The pin may be run through either the pronotum, the scutellum or, in some cases, the right corium (every experienced collector has his or her own preference), but, in any case, care should be taken not to destroy a structural area in the process. This is particularly true of the scutellum in many elongate or ant-mimetic species.

For some orders very small steel headless pins called "minuten nadelen" are pinned into a piece of cork attached to a regular insect pin and the minuten inserted into the right side of the insect's thorax. We have found this method to be generally inferior and more space consuming for hemipterans than point mounting.

European workers use small rectangular cards attached to regular insect pins and glue their specimens onto the cards with the ventral surface down. This introduces an added safety factor against breakage of legs but is really a reprehensible practice as it completely obscures the ventral surface which is often important in identification. It should always be remembered that an insect has two sides but only one top and one bottom.

In addition to the conventional pinning and pointing methods the serious collector will also maintain a collection in liquid preservative. This is essential for the preservation of nymphs and eggs. The former usually shrivel badly when dry mounted. The best preservative is 70% ethyl alcohol. In the field it is often necessary to place specimens directly into alcohol, but superior specimens are obtained by dropping live specimens directly into very hot water, this expands the tissues and aids in the preservation of colors. Specimens desired for future dissection are best preserved in 70% isopropyl alcohol. For chromosome study specimens should be placed in a liquid composed of three parts pure isopropyl alcohol and one part glacial acetic acid. This fixative may be kept for more than three months without losing its effectiveness.

In many cases it is necessary to study the genitalia for critical determination. With a little practice this is generally not difficult but requires care and a pair of very sharp pointed jewelers forceps. To study the genitalia of dry male hemipterans first soften the end of the abdomen by applying to it a relaxing fluid (100% ethanol-100ml., distilled water-75ml., benzol (benzene)-10ml., ethyl acetate-10ml.) with a small camel's hair brush or by placing the end of the abdomen in the fluid. When the end of the abdomen has become softened, the genital capsule can be gently teased out. It is then placed in a 10% solution of potassium hydroxide and distilled water and either left overnight as a cold preparation or heated gently for two to three minutes (time involved is variable depending upon the degree of sclerotization of the specimen). The capsule is then removed to distilled water for several minutes, into a second "bath" of distilled water, then into acetic acid for two to three minutes, and then to 70% ethyl alcohol. It then is transferred to glycerin where it may be dissected and studied. With females it is usually best to remove the entire abdomen and then follow the same procedure as for males.

Vials for preserving insects in liquid vary with the size of specimens and the individual taste of the collector. However, in general, it is wise to adopt a standard size for ease of storage. We prefer either a four or six dram lip vial. This size will be sufficient for all but the very largest specimens and will provide sufficient liquid for specimens to be retained for a long time without drying out. To conserve space small specimens may be placed in small shell vials, plugged with cotton and several placed in each larger vial. Neoprene stoppers should be used if possible. Large shell vials should be avoided like the plague as they do not close tightly and material dries out in a very short time. Some collectors prefer to use slender procaine tubes with neoprene stoppers. These take up relatively little space and have an added advantage since they may be cotton stoppered and placed as a group in large jars of alcohol for permanent storage.

Labeling: Specimens in collections without proper labels have little or no value and great care should be given to this aspect of the preparation. It is possible to neatly print hand labels using a durable rag paper, india ink, and a Rapidograph or crow quill pen. However, this is laborious and time consuming. There are several supply houses that provide attractive small printed labels at nominal cost. Where any considerable number of specimens are to be processed these printed labels are a great convenience. On any label the minimum information should always include the collection locality, the state, and/or country, the date of collection and the name of the collector. Date labels should have the month given in Roman numerals, or as an abbreviation of the month. Thus October 11, 1976 would read either X-11-1976 or Oct. 11, 1976. This sequence is essential because specimens are often exchanged with collectors in other countries. In Europe the day is conventionally given before the month, thus a label 10-11-1976 would mean October 11 in the United States but November 10 in Europe. In addition to this basic label the value of the specimen is greatly enhanced if additional data on habitat can be added. If the plant species from which the insect was obtained is known this should always be included. In mountainous country it is extremely important to include the elevation. General habitat notes such as "forest floor, marsh, wet meadow, sand dunes, saline lake, etc." also are often of much value. To conserve space in mounted collections it is preferable, and also more attractive, to place the data on two or three small labels rather than a single large one. On point mounted specimens the label should be pinned through its right end so that the major part of the label lies directly below the specimen. For specimens pinned through the body the long axis of the label should follow the long axis of the insect's body with the label pinned so that as little of the label extends on either end of the specimen's body as possible. The use of a "step block" with graduated heights for the point and various labels adds to the uniformity and attractiveness of the collection. While "step blocks" may be purchased from a supply house or easily made with a single hole for each level a more efficient method is to take a thick unwanted book and cut it transversely into pieces of different heights. When these pieces are placed upright in a tight wooden or metal holder you will have a "block" that can handle many labels at a time without the necessity of handling each label individually. For specimens in alcohol the same labels may be used, but more efficient labels are printed by hand with India ink and cut into sizes slightly shorter than the length of the vial and placed directly into the liquid. Never glue labels to the outside of the vial, in time they are certain to be either destroyed or rubbed so badly that the data becomes illegible.

Housing the collection: Most elementary textbooks give adequate discussions for housing collections. Our experience has been that you should use the most expensive method you can possibly afford. There is nothing more discouraging than to find that after a year or two one's collection has been eaten by dermestid beetles. However, if sufficient care is taken any closed box or container can be used. We prefer commercially made, glass topped, tight fitting trays housed in a metal case. These are expensive, running $200-300 for a case of twelve trays. A handy carpenter can however duplicate these at a fraction of the cost. The wooden "Schmidt box" is also excellent if made carefully at home, but has become so expensive commercially as to make the use of these questionable. If glass topped trays are used cardboard units of various dimensions with a styrofoam or similar bottom are a great convenience. Beginning collectors can use cigar boxes with cardboard bottoms but these must be watched carefully for dermestid damage. We emphasize again the desirability of buying or making the very best housing for the collection you can afford. Entomology is one of the scientific areas where the enthusiastic amateur can still make important contributions, and valuable specimens, painstakingly collected and prepared, should merit careful, safe maintenance.

The Workers

No less interesting than the insects themselves to the student of the true bugs are the people who study them. It is always a source of satisfaction and wonder to see the diversity of human beings brought together in a common intellectual and enthusiastic interest for the always absorbing and often perplexing world of the bugs. It is of course impossible to mention all of the people who have made important contributions to what we know of these insects but like other areas of science they include the brilliant, the thorough, the superficial, the neophyte, and the master. Hemipterologists have included men of the highest intellectual attainments and men of most limited vision, but never, we think, have they included the charlatan or the renegade and almost without exception they have been enthusiasts.

In this short section we would like to introduce some of the Hemipterologists of the past. These are the people who have made possible whatever merit this little book has. If it in a small way helps to lead some of you to a place among their number they will, we are sure, feel more than amply repaid.

THE EUROPEANS

Much of the early work on American Hemiptera was accomplished when European science dominated the entire world. Among the greatest of the European Hemipterists are the following:

Bergroth, Ewald. A Finnish doctor, the conscience of his generation, who flayed with acid tongue the careless and superficial (in several languages). A brilliant scientist unparalleled for command of the literature and for his careful and wide-ranging work.

China, William E. The present emeritus keeper of insects at the British Museum. A careful worker who has produced a body of work central to concepts of the evolution and phylogeny of the Hemiptera but whose scientific importance is perhaps best memorialized in the work of almost every current student whom he has helped with decades of advice, aid, and encouragement.

Distant, William Lucas. A dedicated worker, often superficial, always controversial who, nevertheless, left behind him a body of superbly illustrated work of the very greatest importance to every serious student.

Horvath, Geza. The Hungarian "alt Meister" who in his nineties continued to produce outstanding studies of many families of Heteroptera. One of the very few humans who ever approached attainment of a knowledge of the world fauna of Hemiptera.

Kirkaldy, George Willis. A brilliant and controversial Scotchman, for many years resident in Hawaii. Author of a world catalogue of the Pentatomoidea and a major student of the phylogeny and evolution of the Heteroptera.

Reuter, Odo Morannal. A Finnish professor who produced over 500 papers on the Hemiptera, many of them exhaustive monographic studies meticulously constructed and analyzed. The greatest student of the enormous and complex family Miridae, and who, while blind, constructed the first modern phylogenetic analysis of the Heteroptera.

Stål, Carl. A Swedish worker, possibly the greatest of all Hemipterists, who in the short period from the 1850's to the 1870's produced some of the most fundamental studies that have ever been done. An almost intuitive genius so often correct that a good rule of thumb to be adopted is "if you disagree with Stål look twice again."

THE CLASSIC AMERICANS

Barber, Harry Gardner. A New York school teacher and National Museum curator. For decades the outstanding authority on the complex family Lygaeidae but whose wide ranging interests encompassed a large number of heteropterous families. A charming man remembered by many current workers for his kindness, encouragement, and unfailing helpfulness.

Blatchley, Willis Stanley. One of the last of the great, old, field naturalists and best remembered for his comprehensive manuals on Heteroptera, Coleoptera, and Orthoptera. A man of encyclopedic interests combined with the freshness of the dedicated amateur to make him one of the most famous and unique of our faunistic entomologists. His "Heteroptera of Eastern North America" remains the "Bible" for all beginning North American students of the Heteroptera.

Drake, Carl John. An Iowa State University professor. The leading world authority on the Tingidae. A prodigious worker who contributed to knowledge of many families, frequently in a controversial but always stimulating manner.

Hungerford, Herbert Barker. A Kansas University professor. For a lifetime the leading student on aquatic and semi-aquatic Hemiptera. His ability to produce comprehensive and definitive works and his stimulation of a large number of students have left a lasting mark on the evolution of knowledge of the Heteroptera.

Hussey, Roland F. A University of Florida professor who worked on many families of terrestrial and aquatic bugs. A keen student of bibliography and nomenclature and producer of the definitive world catalogue of the family Pyrrhocoridae.

Knight, Harry Hazelton. An Iowa State University professor and most famous student of the North American Miridae. The first individual to demonstrate the basic value of the male genitalia as specific criteria in the plant bugs.

McAtee, Waldo Lee. A Washington, D.C. biologist of wide ranging talents known to Hemipterists for his basic contributions to the taxonomy of small, obscure, but phylogenetically important families and for revisional studies of great value in many diverse families.

Parshley, Howard Madison. A Smith College professor. A man of great intellectual depth best remembered for his indispensable "Bibliography of North American Heteroptera" and basic contributions to Aradidae and to New England Hemipterology.

Say, Thomas. The "Father" of North American Entomology. Describer of many of our most common and widespread species. Working under incredibly difficult conditions and in an isolation almost impossible for our generation to understand he was able in terse descriptions to abstract the essential characteristics of most species he treated.

Torre-Bueno, José Rollin de la. A New York businessman. Best known for his work on the biology of aquatic and semi-aquatic families, his comprehensive and regrettably incomplete "Synopsis of North American Hemiptera" and for his flights of rhetoric in numerous volumes of the Bulletin of the Brooklyn Entomological Society of which he was editor for many years.

Uhler, Philip Reese. The dean of American Hemipterists and our earliest native student of the first rank. He established much of the basis upon which our present knowledge rests and who first described many of our most common and widespread species.

Usinger, Robert Leslie. A University of California professor of truly catholic knowledge of the Heteroptera who produced encyclopedic and definitive works on the Aradidae and Cimicidae and works of fundamental importance to an incredible number of heteropteran families. A man of advanced conceptual ideas able to bring a wide variety of information together to form his mature taxonomic conclusions. Perhaps the closest approach to Carl Stål that Twentieth Century Hemipterology has produced.

Van Duzee, Edward Payson. A professional librarian in New York and curator at the California Academy of Sciences. A worthy successor to P. R. Uhler, his interest and ability ranged over the entire order. He produced the fundamental catalogue of North American Hemiptera still used by all serious students.

General References

The literature cited below includes only comprehensive general works dealing with the entire order or with a group of families. Extensive papers treating a single family or portion of a family are cited at the end of each family treatment.

Blatchley, W. S., 1926. Heteroptera or true bugs of Eastern North America, with special reference to the faunas of Indiana and Florida. Nature Publishing Company, Indianapolis, Indiana.

Britton, W. E. (edit.), 1923. Guide to the insects of Connecticut. Pt. IV. The Hemiptera or sucking insects of Connecticut. Bull. Conn. Geol. Nat. Hist. Surv. No. 34:807 pp. Hartford.

China, W. E. & N. C. E. Miller, 1959. Checklist and keys to the families and subfamilies of the Hemiptera—Heteroptera. Bull. Brit. Museum (Nat. Hist.) Ent. Series 8:(1):1-45.

Froeschner, R. C., 1941-1962. Contributions to a synopsis of the Hemiptera of Missouri. Pts. I-V. Amer. Midl. Naturalist 26:(1):122-146; 27:(3):591-609; 31:(3):638-683; 42:(1):123-188; 67:(1): 208-240.

Hungerford, H. B., 1920. The biology and ecology of aquatic and semi-aquatic Hemiptera. Kans. Univ. Sci. Bull. 11(21):1-256.

Parshley, H. M., 1925. A bibliography of the North American Hemiptera-Heteroptera. Smith College, Northampton, Mass.

Torre-Bueno, J. R., 1939-1946. A synopsis of the Hemiptera-Heteroptera of America north of Mexico. Entomologica Americana 19:(3): 141-304; 21:(2):41-122; 26:(1-2):1-88.

Usinger, R. L., 1963. Aquatic Hemiptera. In Aquatic insects of California, with keys to North American genera and California species. Univ. of California Press, Berkeley and Los Angeles.

Van Duzee, E. P., 1917. Catalogue of the Hemiptera north of Mexico excepting the Aphididae, Coccidae and Aleurodidae. Univ. California Pubs. Ent. 2:XIV:902 pp.

Key to Adults of North American Families of Hemiptera

1 Compound eyes absent (ectoparasites of bats). Fig. 258 .... (p. 139) POLYCTENIDAE

1a Compound eyes present 2

2 Antennae not visible from above, concealed in grooves or pits on underside of head (chiefly aquatic) 3

2a Antennae distinctly visible from above, not concealed in grooves on underside of head (terrestrial or living on water surface) 9

3 Ocelli present. Fig. 494 .... (p. 243) GELASTOCORIDAE

3a Ocelli absent 4

4 Base of head overlapping anterior margin of pronotum (fig. 481); labium lacking distinct segmentation; fore legs usually scoop-shaped. Fig. 484 .... (p. 230) CORIXIDAE

4a Anterior margin of pronotum overlapping base of head or the two evenly abutting; fore legs of various shapes but never scoop-like; labium distinctly segmented 5

5 Body with elongate respiratory filaments present, extending as an elongate tail-like process from end of abdomen. Fig. 486 .... (p. 238) NEPIDAE

5a Body with at most very short, straplike, terminal abdominal filaments present, never with elongate tail-like filaments present 6

6 Hind tarsi lacking a pair of distinct claws at apex. Fig. 459 .... (p. 224) NOTONECTIDAE

6a Hind tarsi bearing a pair of claws at apex 7

7 Less than 3 mm in length; hemelytra beetle-like, strongly, convexly arched dorsally. Fig. 469 .... (p. 230) PLEIDAE

7a More than 3 mm in length, frequently very large; flattened on dorsal surface .. 8

8   Membranes of hemelytra lacking veins (fig. 493); body usually under 15 mm in length; abdomen lacking short strap-like terminal respiratory filaments .... (p. 242) NAUCORIDAE

8a Membranes of hemelytra with veins (fig. 490); body length usually over 20 mm; short strap-like respiratory filaments present at end of abdomen .... (p. 239) BELOSTOMATIDAE

9   Antennae relatively short and inconspicuous, always shorter than head. Fig. 496 .... (p. 244) OCHTERIDAE

9a Antennae always conspicuous, longer than head 10

10 Body extremely elongate and slender; head longer than thorax including scutellum. Fig. 426 .... (p. 209) HYDROMETRIDAE

10a Body of various shapes, sometimes elongated but in such cases much shorter than thorax and scutellum combined .. 11

11 Claws of at least front tarsus attached well before apex of tarsus (fig. 433); apical tarsal segment cleft 12

11a Claws of all tarsi located at apex of distal tarsal segment 14

12 Middle and hind legs attached nearly as far from each other as middle legs are from front legs. Fig. 438 .... (p. 214) VELIIDAE

12a Middle legs attached much closer to hind legs than to front legs. Fig. 433 13

13 Hind femora greatly exceeding posterior end of abdomen. Fig. 433 .... (p. 210) GERRIDAE

13a Hind femora not or at most very slightly exceeding apex of abdomen. Fig. 440 .... (p. 214) VELIIDAE

14 Antennae 5-segmented 15

14a Antennae 4-segmented 22

15 Size 3 mm or less; hemelytra including clavus largely membranous .... (p. 218) HEBRIDAE

15a Size never less than 4 mm, usually much greater; clavus not membranous 16

16 Labium apparently 3-segmented; prosternum with a median transversely striated groove. Fig. 8 .... (p. 119) REDUVIIDAE

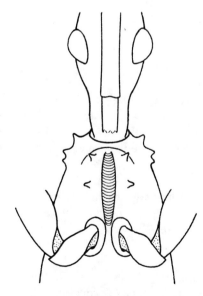

Figure 8

16a Labium distinctly 4-segmented; prosternum lacking a median transversely striated groove 17

17   Tarsi 2-segmented; mesosternum with a prominent median keel .... (p. 40) **PENTATOMIDAE**

17a Tarsi 3-segmented; mesosternum lacking a prominent median keel 18

18   Scutellum small only about 1/5 length of abdomen; a dense pad of hairs present at distal ends of fore and middle tibiae (fig. 9); no trichobothria present on venter of abdomen .... (p. 134) **NABIDAE**

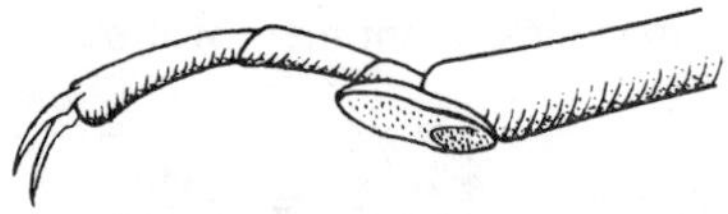

Figure 9

18a Scutellum 1/2 or almost 1/2 as long as abdomen; lacking a dense pad of hairs at distal ends of fore and middle tibiae; trichobothrial hairs present on venter of abdomen 19

19   Scutellum greatly enlarged covering most of corium and attaining or nearly attaining tip of abdomen giving insect a distinct beetle-like aspect. Figs. 41, 54 20

19a Scutellum much smaller, not nearly attaining apex of abdomen and leaving coria broadly exposed. Figs. 28, 71 21

20   Tibiae with 2 rows of thick black spines; small black shining species. Fig. 41 .... (p. 37) **THYREOCORIDAE**

20a Tibiae often with numerous hairs but lacking 2 distinct rows of strong black spines; variously colored but if chiefly black then dull and not brightly shining .... (p. 40) **PENTATOMIDAE**

21   Tibiae armed with 2 or more rows of heavy black or brown spines. Figs. 28, 29 .... (p. 33) **CYDNIDAE**

21a Tibiae with hairs but without rows of heavy spines .... (p. 40) **PENTATOMIDAE**

22   Prosternum with a distinct, finely striated, median longitudinal groove present. Fig. 8 23

22a Prosternum lacking a distinct, striated groove 24

23   Fourth antennal segment clavate (fig. 216); front tarsi hidden in groove of tibiae .... (p. 118) **PHYMATIDAE**

23a Fourth antennal segment slender (fig. 225); front tarsi usually not hidden or reduced .... (p. 119) **REDUVIIDAE**

24   Head divided into separate lobes. Fig. 212 .... (p. 117) **ENICOCEPHALIDAE**

24a Head not divided into distinctly separated lobes 25

25   Scutellum large, nearly completely covering corium, reaching or almost reaching end of abdomen. Fig. 41 .... (p. 37) **THYREOCORIDAE**

25a	Scutellum much smaller, never attaining end of abdomen nor almost covering entire abdomen 26		32	Ostiolar opening of metapleuron present; membrane, when present, distinctly separated from corium. Fig. 267 .... (p. 143) ANTHOCORIDAE
26	Tylus broad, bearing 4-5 distinct tooth-like or peg-like spines .... (p. 33) CYDNIDAE		32a	Metathoracic ostiolar opening absent; wings often meeting in a straight line down middle of back, if not, corium and membrane not clearly separable from one another 33
26a	Tylus not unusually broad, usually without spines but if spines are present never arranged in a peg-like row along anterior margin of a broadened tylus 27		33	Anterior width of scutellum over 1/2 width of posterior margin of pronotum; eyes not overlapping front angles of pronotum .... (p. 206) DIPSOCORIDAE
27	Tips of front and middle tibiae with a thick pad of hairs. Fig. 9 .... (p. 134) NABIDAE		33a	Anterior width of scutellum not over 1/3 width of posterior margin of pronotum; eyes overlapping front angles of pronotum .... (p. 207) SCHIZOPTERIDAE
27a	Tips of front and middle tibiae lacking a thick pad of hairs 28		34	Labium 4-segmented 35
28	Cuneus present on front wings. Fig. 381 29		34a	Labium appearing to be 3-segmented[1] 47
28a	Cuneus lacking on front wings 34		35	Tarsi 3-segmented 36
29	Ocelli absent .... (p. 147) MIRIDAE		35a	Tarsi 2-segmented 44
29a	Ocelli present 30		36	Ocelli absent 37
30	Tarsi 2-segmented. Fig. 275 .... (p. 147) MICROPHYSIDAE		36a	Ocelli present 39
30a	Tarsi 3-segmented 31		37	Metathoracic ostiolar opening present, well developed ...... (p. 70) LYGAEIDAE
31	Labium 4-segmented. Figs. 416, 417 .... (p. 205) ISOMETOPIDAE			
31a	Labium 3-segmented 32			

1. The minute Dipsocoridae and Schizopteridae are keyed here as usually it is very difficult to distinguish four segments.

37a Metathoracic ostiolar opening absent or much reduced 38

38 Pronotum laterally reflexed; seventh sternum entire in female. Fig. 10 .... (p. 95) PYRRHOCORIDAE

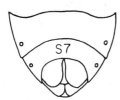

Figure 10

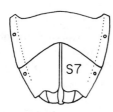

Figure 11

38a Pronotum not reflexed; seventh sternum of female split medially. Fig. 11 .... (p. 96) LARGIDAE

39 Body elongate and slender, at least 8 times as long as maximum pronotal width. Fig. 179 ...... (p. 99) BERYTIDAE

39a Body of various shapes, but if elongate and slender not more than 5 times as long as maximum pronotal width 40

40 Veins of corium with rows of upstanding, slightly recurved spines present. Fig. 181 .... (p. 99) BERYTIDAE

40a Veins of corium lacking rows of upstanding recurved spines 41

41 Membrane with only 4-5 veins (fig. 152)[2]; antennae set below a line drawn longitudinally through middle of eye .... (p. 70) LYGAEIDAE

41a Membrane with numerous (often "intermixing") veins present (fig. 87); antennae set above a longitudinal line drawn through the middle of eye 42

42 Metathoracic ostiolar openings absent or greatly reduced. Fig. 116 .... (p. 66) RHOPALIDAE

42a Metathoracic ostiolar openings large and conspicuous 43

43 Bucculae not extending backward beyond bases of antennae; interocular distance greater than basal width of scutellum. Fig. 108 .... (p. 63) ALYDIDAE

43a Bucculae extending backward beyond bases of antennae; interocular space not as great as basal width of scutellum. Fig. 88 .... (p. 57) COREIDAE

44 Front wings composed of numerous closed (often lacy) cells. Fig. 192 45

44a Front wings not composed of numerous closed cells 46

---

2. Use care in orienting specimen relative to light source so that veins of wing membrane can be readily seen.

45	Ocelli present. Fig. 173 .... (p. 97) **PIESMATIDAE**	50a	Ocelli present 51
45a	Ocelli absent. Fig. 192 .... (p. 106) **TINGIDAE**	51	Eyes and labial segments 1 and 2 bearing conspicuous spines. Fig. 446 .... (p. 219) **LEPTOPODIDAE**
46	Ocelli absent. Fig. 186 .... (p. 103) **ARADIDAE**	51a	Eyes and labium lacking prominent conspicuous spines 52
46a	Ocelli present. Fig. 444 .... (p. 218) **HEBRIDAE**	52	Antennal segments 1 and 2 very short and stout, much less than 1/2 combined length of segments 3 and 4 53
47	Tarsi 2-segmented 48	52a	Combined length of antennal segments 1 and 2 much greater than 1/2 combined length of segments 3 and 4, either segment 1 or 2 (or both) relatively elongate 54
47a	Tarsi 3-segmented 50		
48	Body pale yellow-white, extremely flattened; juga extending anteriorly to reach apex of tylus (fig. 174) (found on Royal Palm) .... (p. 98) **THAUMASTOCORIDAE**	53	Anterior width of scutellum over 1/2 that of posterior margin of pronotum; eyes not overlapping front angles of pronotum .... (p. 206) **DIPSOCORIDAE**
48a	Body usually brown to blackish, not flattened; juga remote from apex of tylus (in ground litter) 49	53a	Anterior width of scutellum not over 1/3 width of posterior margin of pronotum; eyes overlapping front angles of pronotum .... (p. 207) **SCHIZOPTERIDAE**
49	Anterior width of scutellum over 1/2 that of posterior margin of pronotum; eyes not overlapping front angles of pronotum .... (p. 206) **DIPSOCORIDAE**	54	Forewing membrane usually present, composed of 4-5 large cells; eyes very large, kidney shaped. Fig. 451 .... (p. 220) **SALDIDAE**
49a	Anterior width of scutellum not over 1/3 width of posterior margin of pronotum; eyes overlapping front angles of pronotum .... (p. 207) **SCHIZOPTERIDAE**	54a	Forewing membrane (or entire wings) often absent, when present lacking cells and veins, eyes not kidney shaped. Fig. 441 .... (p. 217) **MESOVELIIDAE**
50	Ocelli absent .... (p. 139) **CIMICIDAE**		

Key to Nymphs of North American Families of Hemiptera [1]

1. Compound eyes absent; a comb of spines anteriorly below on head (fig. 12); no dorsal abdominal scent glands present; ectoparasites of bats .... (p. 139) POLYCTENIDAE

Figure 12

1a. Compound eyes present; no comb of spines present ventrally on anterior part of head; dorsal abdominal scent glands usually present 2

2. Antennae short usually concealed below head; if slightly visible from above (Ochteridae) then much shorter than head 3

2a. Antennae conspicuously visible from above, as long as or longer than head 10

3. Dorsal abdominal scent glands present 4

3a. Dorsal abdominal scent glands absent .. 6

4. Scent gland openings present between segments 3-4, 4-5 and 5-6 .... (p. 230) CORIXIDAE

4a. Scent gland openings present between segments 3-4 only 5

5. Scent gland opening single (fig. 13); body strongly arched dorsally .... (p. 230) PLEIDAE

Figure 13

1. Modified from Herring and Ashlock, 1971.

5a Scent gland opening double (fig. 14); body flattened .. (p. 242) NAUCORIDAE

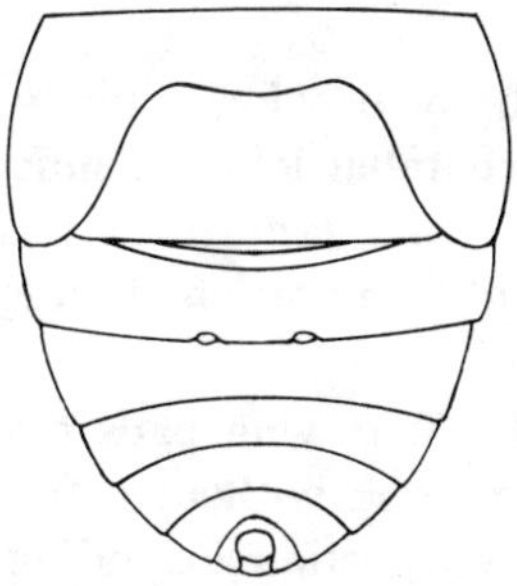

Figure 14

6   Large ciliated metasternal plates present. Fig. 15 .... (p. 239) BELOSTOMATIDAE

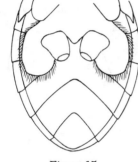

Figure 15

6a Metasternal plates absent 7

7   A pair of long respiratory filaments present on apex of abdomen; body usually long and quite slender. Fig. 486 .... (p. 238) NEPIDAE

7a Respiratory filaments absent; body always elongate-oval or oval and flattened 8

8   Body strongly arched dorsally, elongate-oval .... (p. 224) NOTONECTIDAE

8a Body not strongly arched dorsally, oval and flattened 9

9   Legs slender, fore femora not enlarged; front of head with a crown of heavy upright spines ...... (p. 244) OCHTERIDAE

9a Legs with fore femora enlarged (fig. 495); head without spines .... (p. 243) GELASTOCORIDAE

10 Venter clothed with dense silvery pubescence (semiaquatic) 11

10a Venter with at most scattered silvery hairs present on abdomen 15

11 Dorsal abdominal scent gland(s) present 12

11a Dorsal abdominal scent glands absent 13

12 Antennal segment I short, equal to or shorter than interocular width .... (p. 218) HEBRIDAE

12a Antennal segment I longer than interocular width ...... (p. 217) MESOVELIIDAE

13 Head elongate, longer than thorax; claws apical .... (p. 209) HYDROMETRIDAE

13a Head not longer than thorax; claws subapical 14

Key to Nymphs of North American Families of Hemiptera   27

14 Head with a median, longitudinal suture or groove .... (p. 214) VELIIDAE

14a Head without a median, longitudinal suture or groove .... (p. 210) GERRIDAE

15 Head with 3 pairs of trichobothria on vertex[2] 16

15a Head lacking trichobothria on vertex .. 18

16 Abdomen with dorsal abdominal scent gland openings present between terga 3-4, 4-5 and 5-6 .... (p. 70) LYGAEIDAE

16a Abdomen with dorsal abdominal scent gland openings only between terga 3-4 17

17 Labium and fore femora with many spines .... (p. 219) LEPTOPODIDAE

17a Labium and fore femora without spines .... (p. 220) SALDIDAE

18 Trichobothria present on abdominal venter. Fig. 16 19

Figure 16

18a Trichobothria absent from abdominal venter 29

19 Mesothoracic wing pads widely separated; scutellar lobe prominent, free portion extending posteriorly at least 1/3 length of free portion of wing pad 20

19a Mesothoracic wing pads close together; scutellar lobe without a free portion or with free portion extending posteriorly for at most 1/6 length of free portion of wing pad 22

20 Anterior dorsal abdominal scent gland with a pore at each end (fig. 17), tibiae without strong spines ....
.... (p. 40) PENTATOMIDAE

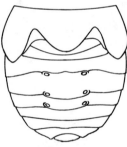

Figure 17

2. The minute bugs of the family Dipsocoridae (less than 2mm in length) often have elongate hairs on the head that can be easily confused with trichobothria.

20a Anterior scent gland without pores (fig. 18); tibiae with strong spines 21

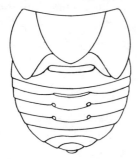

Figure 18

21 An area of non-punctate cuticula located immediately behind eyes on ventral head surface .... (p. 37) THYREOCORIDAE

21a No non-punctate cuticular area present behind head on ventral head surface .... (p. 33) CYDNIDAE

22 Antennae inserted above a line drawn between the center of eye and tip of tylus as seen from side; 2 scent glands present 23

22a Antennae inserted on or below a line drawn between the center of eye and tip of tylus as seen from side; 2 or 3 scent glands present 26

23 First antennal segment filiform, clavate apically, longer than head and pronotum taken together; trichobothria present on abdominal segment 2 only .... (p. 99) BERYTIDAE

23a First antennal segment not filiform nor clavate apically, shorter than head and thorax taken together; trichobothria present on abdominal segments 2 through 6 24

24 Abdominal scent glands lying close to one another so that tergum 5 is constricted at midline. Fig. 19 .... (p. 66) RHOPALIDAE

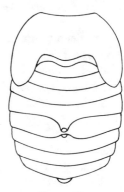

Figure 19

24a Abdominal scent glands not lying close to one another, both slightly displaced posteriorly, tergum 5 not constricted at midline. Fig. 20 25

Figure 20

25 Head including eyes, more than 2/3 and most often nearly equal to width of pronotum; body usually without spines at most with a very few spines present .... (p. 63) ALYDIDAE

25a Head including eyes no more 2/3 width of pronotum; body often covered with spines .... (p. 57) COREIDAE

26   Juga (viewed from above) projecting well in front of tylus; abdominal segments 5 and 6 with a single sublateral trichobothrium on each side .... (p. 97) PIESMATIDAE

26a Juga not projecting in front of tylus; more than one trichobothrium on each side of abdomen placed laterally on some segments, or present medially on segments 2 and 3 27

27   First antennal segment usually equal to or shorter than 1st labial segment but if longer, then at least one pair of spiracles dorsal or only 2 dorsal abdominal scent glands present .... (p. 70) LYGAEIDAE

27a First antennal segment at least 1/3 longer than 1st labial segment; all spiracles ventral; abdomen with 3 dorsal scent glands 28

28   Trichobothrial hairs of 5th abdominal segment arranged in a more or less linear sequence on 3 separate dull spots. Fig. 21 .... (p. 96) LARGIDAE

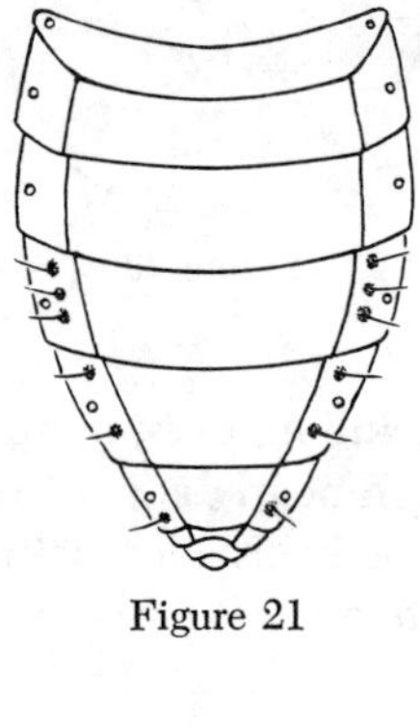

Figure 21

28a Trichobothrial hairs of 5th abdominal segment grouped together at antero-lateral corner on a single dull spot. Fig. 22 .. .. (p. 95) PYRRHOCORIDAE

Figure 22

29   Labium appearing to be 3-segmented 30

29a Labium 4-segmented 37

30   Pronotum with a cross-striated stridulatory groove between fore coxae. Fig. 8 31

30a Pronotum without a stridulatory groove between fore coxae 32

31   Fore femora enlarged, triangular, short; head with a groove above eye for reception of antennae .... (p. 118) PHYMATIDAE

31a Fore femora sometimes enlarged but never triangular; head without a groove above eye for reception of antennae .... (p. 119) REDUVIIDAE

32 Abdomen with many individual pigmented sclerites on a single segment (fig. 23); coiled stylets visible in tylus from beneath, or if not visible, then tylus greatly swollen .... (p. 103) ARADIDAE

Figure 23

32a Abdomen without individual pigmented sclerites on a single segment; stylets not coiled, tylus not greatly swollen 33

33 Wing pads absent .... (p. 139) CIMICIDAE

33a Wing pads present 34

34 Antennal segments 3 and 4 filiform, more than twice length of 1 and 2 taken together 35

34a Antennal segments 3 and 4 often filiform but always less than twice length of 1 and 2 taken together 36

35 Eyes projected outward, not markedly overlapping front angles of pronotum; head and tibiae with strong bristles .... (p. 206) DIPSOCORIDAE

35a Eyes projected both outward and backward, markedly overlapping front angles of pronotum; head and tibiae without strong bristles .... (p. 207) SCHIZOPTERIDAE

36 Tarsi 2-segmented .... (p. 143) ANTHOCORIDAE

36a Tarsi 1-segmented .... (p. 147) MICROPHYSIDAE

37 First labial segment laid open, very wide, stylets plainly visible (Palm Bugs, South Florida only) .... (p. 98) THAUMASTOCORIDAE

37a First labial segment sometimes thicker than following segments but never laid open so that stylets are plainly visible 38

38 With 3 dorsal abdominal scent glands .... (p. 134) NABIDAE

38a With less than 3 dorsal scent glands 39

39 With 2 dorsal abdominal scent glands; body often covered with spines .... (p. 106) TINGIDAE

39a With one dorsal abdominal scent gland; body not covered with spines 40

40 Scent gland on abdominal segment 3; head 2-lobed .... (p. 117) ENICOCEPHALIDAE

40a Scent gland on suture between segments 3 and 4; head not 2-lobed .... (p. 147) MIRIDAE (Including ISOMETOPIDAE)

Key to Nymphs of North American Families of Hemiptera   31

REFERENCES

DeCoursey, R. M. 1971. Keys to the families and subfamilies of the nymphs of North American Hemiptera-Heteroptera. Proc. Ent. Soc. Wash. 73:(4):413-428.

Herring, J. L. & P. D. Ashlock. 1971. A key to the nymphs of the families of Hemiptera (Heteroptera) of America north of Mexico. Fla. Entomologist 54:(3):207-212.

Lawson, F. A. 1959. Identification of the nymphs of common families of Hemiptera. Jour. Kans. Ent. Soc. 32:(2):88-92.

Leston, D. & G. G. E. Scudder. 1956. A key to larvae of the families of British Hemiptera-Heteroptera. Entomologist 89:(1120): 223-231.

Families of Hemiptera

CYDNIDAE
The Burrower Bugs

The Cydnidae comprise a small family of generally black or reddish brown bugs that rarely show any bright coloration. The fore tibiae are flattened, spined and are used for digging into the soil where most of the species live.[1] Burrower bugs are closely related to stink bugs from which they can be distinguished by the fossorial front tibiae, by a series of stout spines on the middle and hind tibiae, and by a fringe of stiff bristles at the tips of the middle and hind coxae. Most species apparently feed on the roots of plants and are not frequently taken by the general collector except when they are attracted to lights.

1   Each clavus meeting beyond tip of scutellum to form a distinct claval commissure. Fig. 24 .... *Amnestus*

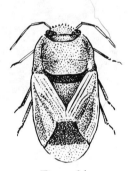

Figure 24

Figure 24   *Amnestus pusillus* Uhler

The members of this genus are the smallest of our burrower bugs (2.5 mm) and are the only ones that possess a distinct claval commissure. *A. pusillus* is yellowish tan in color. The lateral margin of each jugum bears only four stiff spines or pegs. Males have a very long spine present on the underside of the hind femur. Females have a flattened, shining, glabrous median area on the last segment before the genitalia. It is distributed throughout most of the United States and is common at lights.

1. *Scaptocoris castaneus* Perty introduced from South America at Charleston, S.C., where it now appears to be well established, has peculiar falcate fore tibiae.

A. basidentatus Froeschner also has only four spines on each jugum and is colored much like *pusillus* but it is a little smaller, lighter, has a much shorter hind femoral spine and lacks the smooth (glabrous) abdominal area of the female. Its range is over much of the eastern U.S. west to Missouri and Texas.

A. pallidus Uhler is, despite its name, a darker, reddish brown, somewhat larger insect with five spines present on each jugum. The labium does not surpass the posterior coxae. It is widely distributed, apparently almost throughout the United States.

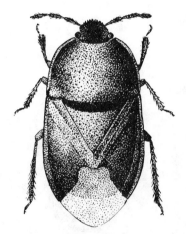

Figure 25

A. spinifrons (Say) (fig. 25) closely resembles *pallidus* but has a longer beak that reaches at least onto the base of the abdomen. It averages slightly larger than *pallidus* (but there is overlap) and is a very rich, red-brown color. This species and *pallidus* are often confused in collections. Males may also be separated by the presence of a sharp angulate expansion of the margin of the fore tibiae opposite the row of strong teeth. It is found in eastern U.S. west to Kansas and Texas.

Two additional species have been taken in the U.S., one at Brownsville, Texas and the other in south Florida.

1a Each clavus not meeting beyond tip of scutellum (fig. 28), thus claval commissure absent 2

2 Body with a distinct white lateral stripe present along margins of pronotum and corium; pronotum lacking a row of setae bearing punctures just within lateral margins .... *Sehirus*

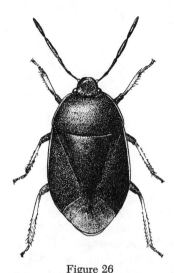

Figure 26

Figure 26 *Sehirus cinctus* (P.B.)

A beautiful shining black or blue-black or dark red-brown insect with a narrow white line running along the lateral margins of the body. The length is 4-7 mm. It is the only burrower bug in N. America with bright contrasting coloration which is frequently taken above the ground level on plants where it feeds frequently on nettles and mints. It has been taken on carrion. It is common throughout the U.S. and southern Canada.

2a Body lacking a white marginal stripe; pronotum usually has a series of setae bearing punctures just within lateral margin 3

3 Pronotum anteriorly with a deep, sharply impressed line paralleling anterior margin to form a distinct anterior collar. Fig. 27 .... *Pangaeus*

Figure 27

Figure 29

Figure 28

Figure 28 *Pangaeus bilineatus* (Say)

This is the commonest large (6-8 mm) black cydnid in most of the eastern U.S. east of the Great Plains, and it occurs in the Southwest to California. It is relatively rare in the northern states (old records from Quebec and New England need verification). The color is strongly shining black; there are 6-8 strong setae (often abraded) on the juga. This burrowing bug has been reported as being destructive to plants in the south. It is attracted by the odor of eugenol or geranol. Little is known of its biology. Four additional species are known from the southern U.S.

3a Pronotum without a deep distinct impressed line behind anterior margin (a row of punctures may occupy this position) .. 4

4 Hind tibiae strongly broadened and flattened (fig. 29), as wide as anterior tibiae, and with spines on dorsal margin much shorter and thicker than those on ventral margin .... *Cyrtomenus*

Cyrtomenus ciliatus (P.B.)

This is a very large shining reddish brown species at once recognizable by the remarkably strongly flattened tibiae on all of the legs. The antennal segments are short and elliptical. The pronotum has an impressed row of punctures running transversely across the middle and has a closely set row of smaller punctures near the anterior margin. It is one of the largest of the N. American cydnids (6-9 mm). It lives on or in the ground near water and occasionally comes to light. It is found in the southern U.S. north to New Jersey and southern Illinois and west to Texas. An additional species occurs in the southwestern states.

4a **Hind tibiae not strongly flattened and not as wide as front tibiae, spines on upper and lower margin nearly equal** **5**

5 Head in front of eyes with a submarginal row of very close set long hairs and short pegs extending from eye to eye. Fig. 30 6

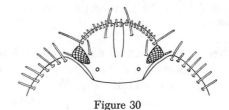

Figure 30

Families of Hemiptera   35

5a Head with only a few widely spaced hairs in submarginal row in front of eye. Fig. 31 7

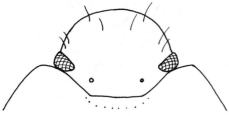

Figure 31

6 Scent gland channel forming a loop surrounding a ventrally visible pore, the outer end broadly rounded. Fig. 32 .... *Microporus*

Figure 32

Microporus obliquus Uhler

M. obliquus (length 4 mm) is a shining reddish brown species. The body is very broadly oval and usually has its maximum width behind the middle. The lateral margins are fringed with elongate hairs, the juga have numerous "pegs" and the scutellum is very broad at the tip. It occurs in sandy places about the roots of plants throughout most of the U.S. An additional species occurs in California.

6a Scent gland channel not forming a loop, but subacute at outer end, pore visible posteriorly, not ventrally. Fig. 33 .... *Tominotus*

Figure 33

Tominotus communis (Uhl.)

This is a dark, polished, nearly black species which can be distinguished from most of our species by the complete row of pegs on the juga. It resembles *Pangaeus bilineatus* in size and shape but lacks the distinct pronotal "collar": length 5-7 mm. It is a southern species that is distributed across the U.S. from Florida to California and north at least to N. Carolina and southern Illinois. Three additional species occur in the southwestern states.

7 Terminal process of scent gland channel flat, expanded posteriorly as a partially polished flap. Fig. 34 .... *Melanaethus*

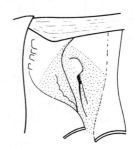

Figure 34

Figure 35

Families of Hemiptera

Figure 36 *Melanaethus pennsylvanicus* (Sign.)

This is a small (3.5 mm) dark shining species, usually widest behind the middle. The head lacks punctures or has only a few scattered ones. The range is southern from Virginia and Florida west to Alabama and Nebraska. Very little is known of its habits.

M. robustus Uhler is a somewhat larger black species readily distinguishable by the coarse thick punctures on the dorsal surface of the head. The range is similar but it is known from as far north as Illinois, Ohio, New Jersey and Pennsylvania. Nine additional species occur north of Mexico. They are chiefly southwestern but three occur in southeastern states and one has been reported from Alaska.

7a Terminal process of scent gland channel neither expanded nor flat. Fig. 35 ....
.... *Dallasiellus*

Dallasiellus discrepans (Uhl.)

This is a medium-sized species (6-8 mm) with the pronotum strongly polished but with the median impression incomplete near the middle, or entirely absent. In general appearance it is very similar to *Tominotus communis*. One must watch closely for the lack of pegs on the inner 1/4 of the jugal margin. It is distributed in the western U.S. from Arizona to Washington but not known for certain to occur naturally east of Oklahoma. (It has been taken once in Philadelphia but possibly was introduced.)

D. lugubris (Stål) is a much smaller species (4-5.5 mm) found in the southern U.S. from Alabama to Texas. Four additional species occur in the southern and western states.

REFERENCE
Froeschner, R. C. 1960. Cydnidae of the Western Hemisphere. Proc. U. S. N. M. 111: 337-680.

THYREOCORIDAE
The Negro Bugs

This is a small family of black, oval, usually shining species that are of small or medium size. The scutellum is very large and covers the entire abdomen. This gives the negro bugs much the appearance of small black beetles and beginning students very frequently confuse them with Coleoptera. The tibiae are spinose as in the burrowing bugs but the front tibiae are not flattened for digging. The various species are usually found on weedy plants in open fields. Most species are very similar in appearance and accurate identification is often exceedingly critical.

1 Lateral edges of pronotum and abdomen with a fringe of long hairs ...... *Cydnoides*

Cydnoides ciliatus Uhler

This is an entirely black species 4 to 5 mm long and usually with a bronzy cast. It is primarily an inhabitant of sandy areas where it burrows at the roots of plants, but it also lives on *Cassia*. The species is widely distributed west of the Mississippi River but it is also found in Florida. The distribution needs further study as it may resemble that of some reptiles and birds in having a disjunct Florida population. There are 5 additional western species.

1a Lateral margins of pronotum and abdomen lacking a fringe of long hairs 2

2 Lateral margins of coria with a distinct longitudinal groove (=costal furrow) just within the edge (fig. 37); corium always completely black .... *Galgupha*

Figure 38

Figure 37

Figure 39

This genus contains 6 species north of Mexico, the majority of which are confined to the southern and southwestern states. The species all closely resemble one another and are difficult to identify accurately without material of other species for comparison.

Galgupha atra A. & S.

This is a large (4.5-6 mm) strongly shining species with very weakly developed punctures on the dorsal surface. The scutellum when viewed laterally is evenly, broadly rounded posteriorly (fig. 38, as opposed to fig. 40). The fore tibiae have the two distal spines on the antero-dorsal row much reduced relative to the others and appearing as slender setae (fig. 39). (To observe this important character it is necessary to orient the single elongate spine on the inner face so that it lies perpendicular to the viewer and then observe the row of spines immediately above.) *G. atra* is a common and widespread species from New England south into Florida and west to Washington, Arizona and Mexico.

G. carinata McAtee & Malloch closely resembles *atra* in size, reduced punctation and in having the two distal spines of the antero-dorsal tibial row reduced to slender setae. It is most reliably separated from *atra* by having the scutellum posteriorly elongately, elliptically tapered rather than broadly, evenly rounded. The distribution is more southern where it occurs from Maryland and Virginia south to Georgia and west to Texas and Oklahoma.

G. aterrima Malloch is a smaller species than the two preceding, seldom exceeding 4.5 mm. It differs in being more conspicuously punctate on the dorsal surface, in not having the two distal spines on the antero-dorsal fore tibial row reduced to setae and when viewed from the side showing a scutellar outline in which the posterior third is sharply angled downward (fig. 40) rather than descending in an even arc. This is a common species, widely distributed from New England west to Wisconsin and south to South Carolina and Oklahoma.

Figure 40

G. *ovalis* Hussey closely resembles *aterrima* in all of the characters listed above for the latter, but is distinguished by having the scutellum when viewed from the side evenly downcurved posteriorly rather than conpicuously angled. It is widely distributed from New England west to Montana and south to Florida and Arizona.

G. *nitiduloides* (Wolff) is an extremely variable species in which the dorsal punctures range from well developed to almost absent. It is frequently confused with *atra* but does not have the distal two spines of the antero-dorsal fore tibial row reduced, is somewhat smaller and more narrowly oval and has the polished area of the metapleuron laterad of the scent gland opening conspicuously punctured. (The four preceding species have this area smooth and polished and at most with only a few scattered punctures). Three subspecies are recognized which range from Oregon to Arizona in the west and across the country into New England, but in the southeast it is not yet reported south of North Carolina.

2a Lateral margins of corium never with a groove running just within the edge, this area usually conspicuously marked with white, yellow or red .... *Corimelaena*

Like *Galgupha* the genus *Corimelaena* contains a considerable number of species north of Mexico (18), and many of them are extremely similar and difficult to identify unless comparative material is available for study.

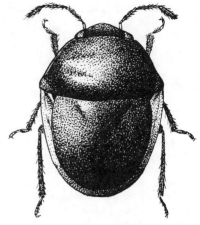

Figure 41

Figure 41 *Corimelaena pulicaria* (Germar)

Throughout most of the country this is the most common of the negro bugs. It often occurs in large numbers on berries where it may cause considerable economic injury. It is a very small (2.5-3.5 mm) black insect readily recognizable by the bluntly rounded corial apex and by having the pale lateral corial stripe expanded basally to extend inward well within the cubital vein (fig. 42) (wing must be pulled laterally from below scutellum for viewing). *C. pulicaria* is distributed over almost the entire country from New England and Ontario south to Florida and west to British Columbia and southwest all the way to Guatemala.

Figure 42

C. lateralis (F.) is also a very common species. It can easily be distinguished from *pulicaria* by being much larger (up to 4.5 mm) and by having the pale corial stripe narrow for its entire length and never broadened basally to extend within the cubital vein. (Some recent literature refers to this species under the name *gillettii* V.D., a junior synonym.) It is a widespread species from New England south to Florida and west to North Dakota and Texas.

C. nigra Dallas is the only widespread species of *Corimelaena* that has the corium entirely black. Care must therefore be taken not to confuse it with species of *Galgupha*. The apex of the corium is broadly rounded. It is a relatively large species for the genus (4.5-4.75 mm), widely distributed in the western states and has been taken in Michigan and New York.

C. marginella Dallas is a small species like *pulicaria* but has the pale corial stripe confined to an area completely laterad of the cubital vein and has the apex of the corium pointed rather than bluntly rounded (fig. 43). *Marginella* is distributed from Rhode Island south through Florida and west to Texas and Kansas. It is a common species in the southeastern states.

Figure 43

In the western states three widely distributed species occur which differ from those discussed above in that the spiracles on abdominal segments three to six are located in the lateral abdominal marginal carina (subgenus *Parapora*) rather than below the carina. These three species can effectively be distinguished only by details of the dissected male genital capsule (*C. incognita* M. & M., *C. virilis* M. & M., *C. extensa* Uhler).

REFERENCE
McAtee, W. L. & J. R. Malloch 1933. Revision of the subfamily Thyreocorinae of the Pentatomidae (Hemiptera-Heteroptera). Ann. Carnegie Mus. 21:191-411.

PENTATOMIDAE
The Stink Bugs

These are the true "stink bugs." They are generally oval or elliptical in shape and are moderate to large in size. The antennae are five segmented; ocelli are present; the head is usually tapering and considerably narrower than the maximum width of the pronotum. The majority of species are plant feeders and they live above ground level on their host plants. Some species become destructive to cultivated plants. One of the subfamilies is predaceous and several of its members are beneficial as important predators on destructive insects. This is a very large family with at least 76 genera known to occur north of Mexico.

The Pentatomidae as treated here is frequently segregated into separate families. In particular those genera with very large scutella which cover or nearly cover the abdomen are treated in most American literature as the family Scutelleridae. The genera *Elasmucha* and *Elasmostethus* have by some recent authors been placed in the family Acanthosomidae or Acanthosomatidae. The turtle bugs (*Amaurochrous*) are sometimes listed under the family Podopidae.

1   Tarsi two segmented 2

1a Tarsi three segmented 3

2   Hind margin of pronotum laterad of base of scutellum bent strongly downward (depressed) and produced backward; scent gland channel short, lobed (ear-like) reaching not more than half way to lateral margin. Fig. 44 .... *Elasmucha*

Figure 44

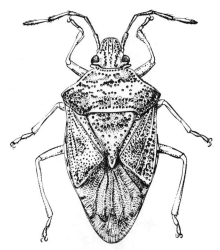

Figure 45

Figure 45   *Elasmucha lateralis* (Say)

This species is a yellowish-green color which in dead specimens fades to dull yellow. The body surface is marked with patches of reddish brown or reddish yellow with conspicuous coarse large brown patches (length 7-9 mm). It is found most frequently on white birch (*Betula alba*). This insect has the remarkable habit of providing maternal care. The female remains with the nymphs and appears to protect them until they are more than half grown. The distribution is northern, from Quebec and New England west through southern Canada and the northern states to the west coast. Much American literature treats this species under the generic name *Meadorus*.

2a Postero-lateral pronotal margins not strongly depressed, area immediately outside base of scutellum not produced backward; scent gland channel elongate, slender, not lobed, extending to very near anterior margin of sclerite and much more than half way to lateral margin. Fig. 46 .... *Elasmostethus*

Figure 46

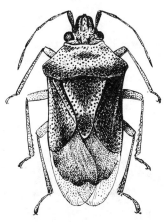

Figure 47

Figure 47   *Elasmostethus cruciatus* (Say)

E. cruciatus is a dull greenish-yellow species with reddish markings at the base of the pronotum and scutellum, at the apex of the corium, on the clavus and on the adjacent area of the corium. The specific name is apparently based on the red "X" which is formed by the clavus and adjacent inner corial stripe appearing diagonally continuous along the apical margin of the opposing corium. The pronotum has widely separated but prominent punctures. Length 10-11 mm. This is a northern species that ranges across southern Canada and the northern tier of states from coast to coast. There are scattered southern records, but these probably are from higher elevations.

E. atricornis (V. D.) can readily be recognized by having black rather than pale antennae, and by having relatively small, fine pronotal punctures. It breeds on the spikenard (*Aralia racemosa*) and appears to be widely distributed although it is usually rare. There is a third species in Alaska and northwest Canada.

sedges in damp places, and sometimes comes to lights in numbers. The species is distributed over much of the United States east of the Great Plains.

A. brevitylus Barber and Sailer is a species of northern distribution occurring from New England west to Nebraska and Kansas. It may be distinguished from *cinctipes* by having the juga exceeding the tylus and sometimes meeting in front of it, instead of only reaching the tip of the tylus as is the case in *cinctipes*.

Three additional species occur in the southeastern states and a fourth is known from California. These insects are known as turtle bugs and are related to five additional scarce genera found in the southern and southwestern states.

3   Scutellum greatly enlarged, usually attaining or nearly attaining posterior end of abdomen (fig. 48) and covering greater part of fore wing 4

3a Scutellum much smaller, usually triangular, remote from posterior end of abdomen and with fore wing broadly exposed laterally 9

4   Antero-lateral pronotal angles distinctly tuberculate (fig. 48); dark yellow-brown or blackish in color .... *Amaurochrous*

4a Antero-lateral pronotal angles not distinctly tuberculate, or if so then color not uniformly dull brown; color variable 5

5   Color variegated with dark blue and either white, orange-yellow or red .... *Stiretrus*

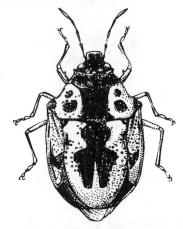

Figure 49

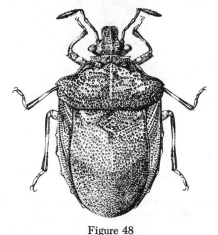

Figure 48

Figure 48 *Amaurochrous cinctipes* (Say)

This is a dark brown insect with the corner of each anterior pronotal angle bearing a short sharp tooth. It is usually found at the bases of grasses and

Figure 49 *Stiretrus anchorago* (Fabr.)

This is a large (8-11 mm) beautiful blue and white (or yellow, orange and red) insect. The color markings are variable but usually include a broad dark central pronotal stripe with a large pale area posteriorly on either side, the pale patch containing one to three dark spots. It feeds on the larvae of Coleoptera and Lepidoptera. The range is over most of the eastern United States and west to the Great Plains.

5a Color variable, most frequently brown, yellow or dull black, never variegated with blue and either white, orange-yellow or red 6

6 Abdomen below lacking a stridulatory area on either side of the fourth and fifth segments; corium exposed laterad of scutellum for nearly its entire length .... *Eurygaster*

Figure 51

7 Pronotum with a transverse groove or impressed area near the middle; under 6 mm in length .... *Acantholomidea*

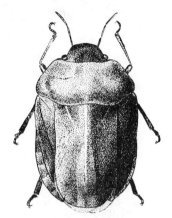

Figure 50

Figure 50 *Eurygaster alternatus* (Say)

This is a yellowish-brown insect with the scutellum narrower than in most other related species, leaving the coria exposed outside the scutellum for nearly their entire length. The scutellum usually has a pale median stripe (length 6.5-10 mm). It is usually found on sedges and grasses in damp meadows. The distribution is through the northern states from coast to coast. It is rare or absent south of Maryland in the east, but has been reported from Arizona and California in the west. Four species are currently recognized as occurring in North America.

6a Fourth and fifth abdominal segments below with a series of very fine grooves and ridges (stridulatory area) on either side of midline (fig. 51); corium exposed laterad of scutellum only on anterior half 7

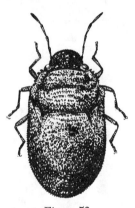

Figure 52

Figure 52 *Acantholomidea denticulata* Stål

This is a small dull black stink bug, usually with white markings on the outer basal angles of the lateral portion of the abdomen, on each side of the scutellum near its base and as a thin streak medially on the outer half of the scutellum. The body is strongly punctured with each puncture bearing a yellow hair. The lateral margins of the pronotum have a number of small teeth. It is a scarce little species which breeds on the seeds of *Ceanothus pubescens* (T. & G.). Its distribution is poorly understood. It is known from Connecticut and New York west to Kansas but has not been reported south of New Jersey and Illinois.

An additional species which occurs in the southern states is very similar to *denticulata* in size and color but has the lateral pronotal margins entire rather than denticulate.

7a Pronotum evenly convex, lacking a transverse impression 8

8   Length 12 mm or more; scent gland opening short, slightly curved forward, but not continuing outward as a long channel. Fig. 53 .... *Tetyra*

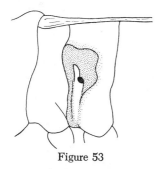

Figure 53

8a Length 4-8 mm; scent gland channel elongate, broadened and abruptly curved anteriorly at distal end. Fig. 55 .... *Homaemus*

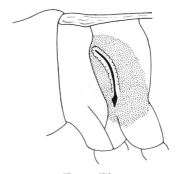

Figure 55

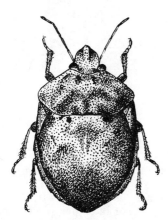

Figure 54

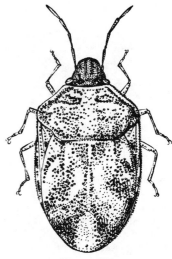

Figure 56

Figure 54 *Tetyra bipunctata* (H. S.)

This is one of our largest shield bugs (12-17 mm). It is brownish yellow in color, interspersed with brown to black punctures and spots, and it usually has a small dark spot on either side of the scutellum near the base which gives the insect its name. It is a scarce species usually found on pine trees. The range is from New York and Indiana south through Florida. Three additional species occur in the southern and southwestern states.

Figure 56 *Homaemus aeneifrons* (Say)

In males the coloration is dull yellowish becoming a mottled gray. Females are usually a brighter yellow with irregular darker patches, a distinct pale median line on the posterior third of the scutellum and with the head bronze-colored without pale margins (length 7-9 mm). It is usually found in wet meadows on grasses and sedges. The range is from eastern Canada south to Maryland and westward to the Pacific Coast. In the west it occurs south to Arizona and California.

H. bijugis is a closely related species, differing from *aeneifrons* by having a pale line running longitudinally near the margin of each jugum. The range is western over much of the country west of the Mississippi and eastward into Illinois.

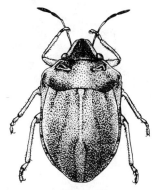

Figure 57

H. parvulus (Germar) (fig. 57) is a smaller species (4-6 mm), similarly colored but generally of a paler yellow with the scent gland channel evenly curved toward the tip. It lives on grasses in sandy places and in the southern states is one of our commonest and smallest shield bugs. The range is southern from North Carolina south to Florida and from southern Illinois westward to Colorado, Arizona and California. Two additional species occur in the Southwest.

9 First segment of labium freely exposed, not lying in a groove formed by bucculae; bucculae coming together posteriorly on underside of head (fig. 58) (predaceous stink bugs) 10

Figure 58

9a First segment of labium lying in a groove between bucculae so that it is often obscured and the beak often has appearance of not beginning at front of head; bucculae parallel and not coming together posteriorly on underside of head (fig. 59) (plant-feeding stink bugs) 12

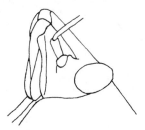

Figure 59

10 Front femur with a small tubercle or spine below on distal third (sometimes almost obsolete, if so, colors bright red and black); usually marked with bright colors —red, white, yellow or orange .... *Perillus*

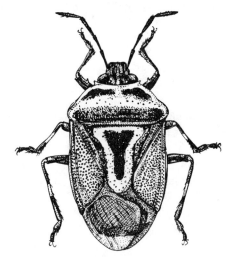

Figure 60

Families of Hemiptera 45

Figure 60 *Perillus bioculatus* (Fabr.)

This is a brightly marked stink bug with the scutellum, corium and thorax usually margined with red, yellow, white or orange (8.5-12 mm). The color is extremely variable and depends primarily upon the temperatures at which the bug develops. This stink bug is an important predator upon the Colorado potato beetle, and has been introduced into Europe in an attempt to control the damage caused there by the beetle. It is widely distributed in the United States.

P. circumcinctus Stål occurs in the northern states from New England to the Dakotas and south to New Jersey and Missouri. It can be separated from *bioculatus* by the lack of a row of black abdominal spots, by not having the abdominal spiracles enclosed in black and by having the first two and the proximal one half of the third antennal segment reddish brown. It is similar to *bioculatus* in size.

P. exaptus (Say) is a widespread but scarce species in the United States, very variable in color but recognizable by having only a blunt tubercle on the fore femur below instead of a stout spine, and this so small as to be barely visible in some individuals. It is much smaller than *bioculatus* (5.75 mm), and does not have each spiracle enclosed in a black spot. Three additional species are known from the western states.

10a Front femur unarmed below; usually brown or gray-colored species 11

11 Smaller species, usually considerably less than 14 mm in length; abdomen below with one or two series of small black spots running midway between midline and lateral margins .... *Podisus*

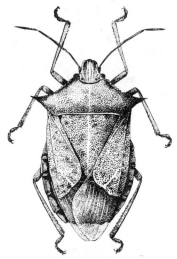

Figure 61

Figure 61 *Podisus maculiventris* (Say)

This is the most common predatory stink bug throughout much of the country. It is of a nearly uniform brown color, frequently marked with rose-red distally on the corium, with sharply pointed humeral angles. The abdomen has a spine medially at the base which projects forward between the hind coxae. Each wing membrane usually has a dark longitudinal median stripe at least near the tip (length 6-8 mm). Beginning students frequently confuse *maculiventris* with members of the similarly colored plant-feeding genus *Euschistus*.

P. serieventris Uhler is very similar to *maculiventris* in appearance but has a much shorter ventral abdominal spine which does not reach forward to the hind coxae, and it has a darker ground color and obtuse rather than sharply acute humeral angles. The range is northern across southern Canada and the northern states from New England to the Pacific Coast.

P. placidus Uhler is found in the northern states at least as far west as Colorado. It has the humeral pronotal angles rounded instead of sharply pointed, and lacks the dark stripe on the wing membrane which is so characteristic of *maculiventris* and *serieventris*. This species is commonly found in webs of the fall webworm (*Hypantria cunea*) and the tent caterpillar (*Malacosoma*).

P. modestus (Dallas) closely resembles *maculiventris* in size and color, but has a much shorter abdominal spine which does not extend well between the posterior coxae. The humeral angles, while sometimes acute, are not produced into a distinct sharp spine. This is a northern species occurring across the entire country but rarely taken south of 38° latitude.

There are at least six other species known from the United States, most of them southern in distribution.

11a Larger species, usually more than 14 mm in length; abdomen below lacking a row of black spots .... ***Apateticus***

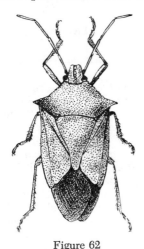

Figure 62

Figure 62 *Apateticus cynicus* Say

A very large (17-20 mm) uniformly bright tan or clay yellow-colored insect with sharply pointed pronotal humeri and with the juga extending slightly beyond the tylus. This is a very widespread but usually uncommon species occurring from eastern Canada over much of the United States west to the Dakotas and in the South to Arizona. *A. cynicus* is a predator of a variety of caterpillars. Five additional closely related species are known.

12 Juga with teeth on outer margin near tip .... ***Brochymena***

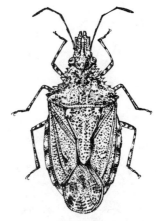

Figure 63

Figure 63 *Brochymena quadripustulata* (Fabr.)

This is a dark reddish-brown or grayish insect with the surface irregularly marked with small white and dark areas so that it resembles tree bark. The juga greatly exceed the tylus and usually come together in front of it. The humeral pronotal angles are bluntly pointed, and the scent gland canal has a distinct black ear-like distal end. It is the most common member of the genus and is distributed over the entire United States although most abundant north of 40° latitude.

B. sulcata V. D. closely resembles *quadripustulata* but differs in not having the juga in contact and overlapping mesally in front of the anterior end of the tylus, but leaving a narrow open space between the inner margins of the juga. It has a lighter, more "ashy" color than does *quadripustulata*, sharper humeral angles and males have a prominent sulcus across the apical edge of the genital capsule. This is a very common species in the southwestern states where it largely replaces *quadripustulata*.

B. parva Ruckes differs from *quadripustulata* in having the juga at most only slightly exceeding the tip of the tylus. It is found in the southwestern states.

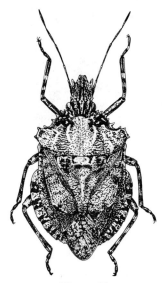

Figure 64

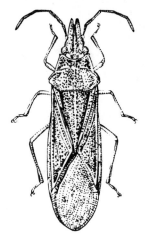

Figure 65

Figure 65   *Mecidea major* Sailer

B. *arborea* (Say) (fig. 64) has the humeral pronotal angles squared off (subquadrate) **rather** than bluntly pointed, and lacks an ear-like lobe on the scent gland canal. It is a common species in eastern North America west to the Great Plains. There are 19 additional species known from the United States.

12a   Juga smooth, lacking teeth along outer margin 13

13   Body elongate and slender, nearly linear, about 4 times as long as broad .... *Mecidea*

This is the most elongately slender of all of the North American stink bugs (9-12 mm long). The color is pale straw yellow with the juga surpassing and frequently meeting in front of the tip of the tylus. *M. major* feeds primarily on grasses, particularly *Bouteloua curtipendula* (Michx.) Torr., and occurs from Texas where it is most common north to Kansas and Missouri.

M. minor is somewhat smaller and may be distinguished by having an obscure dark stripe running along the midline of the ventral surface of the abdomen. It is distributed from central California and Utah east to Kansas and Texas.

13a   Body relatively broad and ovoid, not over 3 times as long as broad 14

14   Base of ventral surface of abdomen with a forward projecting spine or tubercle in the middle which reaches between or beyond the hind coxae. Figs. 66, 67 15

14a   Base of abdomen lacking a median forward projecting spine or tubercle 18

15 Ventral spine at base of abdomen very elongate, compressed, reaching forward between the middle coxae. Fig. 66 .... *Piezodorus*

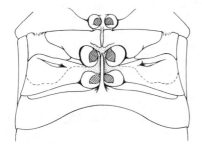

Figure 66

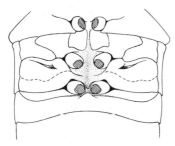

Figure 67

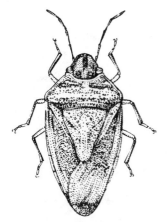

Figure 68

Figure 68 *Piezodorus guildini* (Westwood)

This is a nearly uniformly delicate pale green insect which has the abdominal spiracles a conspicuously contrasting black, and sometimes is marked with red or purple on the posterior portion of the pronotum. There is usually a pale or red smooth calloused band transversely across the center of the pronotum. When the insect is dead the green quickly fades to dull yellow. It is primarily a Neotropical species and has been found only in the extreme southern United States and is common in parts of Florida. It has been reported from as far west as New Mexico.

15a Ventral abdominal spine not projecting as far forward as middle coxae, sometimes represented only as a broad tubercle or projection. Fig. 67 16

16 Second antennal segment half or less than half-length of segment 5; length 12 mm or less; color frequently green but often with darker markings .... *Banasa*

Figure 69

Figure 69 *Banasa dimidiata* (Say)

This insect is usually a bright green color with dark reddish markings across the posterior half of the pronotum and over the corium. The coloration is variable, often with much purplish or reddish brown replacing the green of the scutellum and wings. Sometimes it is nearly uniformly green or even chiefly pale tan. It is cryptically colored and often found on berries. It often comes to lights. The range is over most of the United States.

B. calva (Say) closely resembles *dimidiata* in color, but is usually slightly larger (11-12 mm as compared to 8.5-11 for *dimidiata*). The green color is usually paler and of a more translucent quality. It can be distinguished by having large black spots ventrally on the postero-lateral angle of each abdominal segment, whereas in *dimidiata* these spots are very minute or absent, and by having a longer second antennal segment (3/4 as long as segment 3; in *dimidiata* the ratio is 1:2). It is widely distributed across the continent in the northern states, but with only scattered records in the South. Six or seven additional species are known from the United States.

16a Second antennal segment more than half the length of segment 5; length 14 mm or greater; color a rich green 17

17 Scent gland channel short with tip rather abruptly rounded (subtruncate), channel thus reaching less than 1/3 distance to lateral margin of metapleuron. Fig. 70 .... **Nezara**

Figure 70

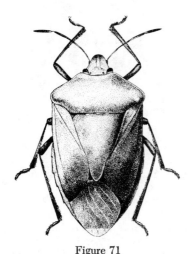

Figure 71

Figure 71 *Nezara viridula* (L.)

This species and members of the genus *Acrosternum* are very similar appearing large, usually uniformly bright green stink bugs. *Viridula* is probably an introduced species and is common only in the southeastern states west to Texas. It frequently does considerable damage to various truck crops throughout the subtropical regions of the world. There are a number of distinctly marked color varieties. In one of these the anterior third of the pronotum and the anterior half of the head are pale yellow or white.

17a Scent gland channel elongate, coming to an elongate tapering point, reaching more than halfway to lateral metapleural margin. Fig. 72 ...... **Acrosternum**

Figure 72

Acrosternum hilare (Say)

A. hilare is an elongate-oval bright green insect with the lateral pronotal margin narrowing anteriorly in nearly a straight line rather than a convex arc (fig. 73), and with the juga equaling the tylus in length. The head and pronotum frequently have a narrow orange or yellow lateral margin. It is a common species along woodland margins and adjacent fields and sometimes becomes destructive to orchard fruits. It is distributed almost throughout the United States.

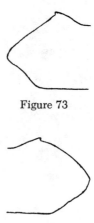

Figure 73

Figure 74

A. pennsylvanicum (De Geer) is also a bright green species which may be distinguished by its more oval shape and especially by having the lateral pronotal margins strongly arcuately rounded (fig. 74). It is widely distributed from Quebec west to Iowa and south to Florida, but is rare in the North. In Florida it appears to breed only on the bracken fern *Pteridium aquilinum* (L). Kuhn. spp. *caudatum* (L.) Bonap.

A third species occurs in southern Florida.

18   Dorsal surface chiefly or entirely green 19

18a Dorsal surface variously colored but never with extensive green coloration 20

19   Metathoracic scent gland auricle terminating in a short tapering lobe, not continued laterally into an elongate nonelevated groove; large species, usually over 11 mm in length .... *Pitedia*

the scutellum. The species is widely distributed from the Great Plains west, and is very common in some of the southwestern states. "Say's stink bug" and related species are often economically destructive to cotton and many other crops.

P. ligata (Say) is known as the "conchuela." It is very similar to *sayi* but lacks the three conspicuous pale dots at the base of the scutellum. It also is a western species.

P. persimilis Horvath lacks the small white spots on the dorsal surface of the body. It is the only *Pitedia* found in the eastern states and it is not usually common except in sandy dry areas.

The status of the various species of *Pitedia* is not clearly understood, and there appears to be considerable hybridization. A careful study of the variability of this genus offers an attractive problem to the serious student. Much of the past literature used the generic name *Chlorochroa*.

19a Metathoracic scent gland opening terminating in an elongate, strongly tapering groove; smaller species, usually under 11 mm in length .... *Thyanta*

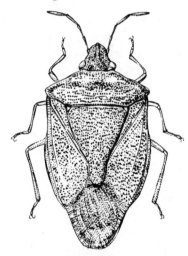

Figure 76

Figure 76 *Thyanta accerra* McAtee[2]

This is a generally nearly uniformly pale green stink bug with angulate or slightly spinose humer-

> 2. *Thyanta accerra* has until very recently been considered as a subspecies of *pallidovirens* Stål. However, it has recently been shown that the two have different numbers of chromosomes, and *pallidovirens* must properly be restricted to the far western states.

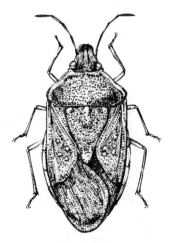

Figure 75

Figure 75 *Pitedia sayi* (Stål)

This is a bright green elongate-oval species with numerous smooth white spots, "points" or irregularities scattered over the dorsal surface, and with three conspicuous white spots across the base of

al angles. The lateral pronotal margins are often narrowly edged with red. It is an extremely common species throughout almost the entire United States east of the Rocky Mountains with a closely related species present in the West. Autumnal populations are sometimes very different in appearance, being extensively marked with brownish or olivaceous with a pale longitudinally raised scutellar line and small pale "dots" scattered over the dorsal surface.

T. custator (F). differs in possessing a dark margin to the antero-lateral areas of the pronotum and in having two series of small black spots on the abdominal venter. This species is restricted to the eastern coastal plain from New England to Florida and along the Gulf Coast to Texas.

Approximately 12 additional species occur north of Mexico.

20   Dorsal surface brightly colored, with conspicuous red and black or orange and black markings 21

20a Dorsal surface dull colored, usually brown, sometimes nearly black with ivory white markings but never conspicuously marked with red and black or orange and black coloration 22

21   Metathoracic scent gland orifice ending in a short rounded ear-like lobe .... *Cosmopepla*

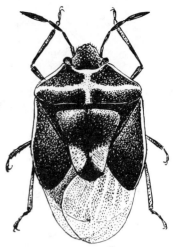

Figure 77

Figure 77   *Cosmopepla bimaculata* (Thomas)

This is a strikingly marked shining black insect (5-7 mm in length) with bright red or yellowish stripes on the pronotum narrowly along the lateral margins, broadly across the humeri and anteriorly down the median line. The scutellum has a large red spot present on either side near the tip. It is common over most of the United States and is reported from a bewilderingly diverse group of host plants.

Four additional species are found in the western states.

21a Metathoracic scent gland orifice ending in a long slender tapering point .... *Murgantia*

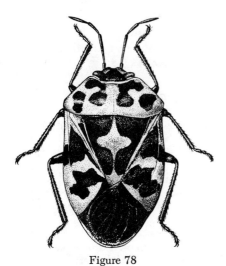

Figure 78

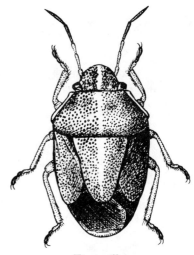

Figure 79

Figure 78 *Murgantia histrionica* (Hahn)

The bright variegated orange and black markings at once distinguish this beautiful insect. Unfortunately, it is a very destructive species in the southern United States where it is known as the harlequin bug. It causes much injury to cabbage and related crops. In years of abundance it spreads northward as far as New England.

One additional species is known from the Florida Keys.

22   Lateral corial margins extending very little if at all further toward tip of abdomen than does scutellum; scutellum evenly and only slightly tapering, its tip rather broadly rounded. Figs. 79, 81 23

22a Lateral margins of coria extending much further toward apex of abdomen than does scutellum (fig. 82); scutellum usually rather abruptly narrowed on apical third with tip narrow 24

23   Juga extending in front of and meeting before tip of tylus .... *Neottiglossa*

Figure 79 *Neottiglossa undata* (Say)

This is one of the smallest North American stink bugs (4.5-5 mm) and rather elongate-oval in shape. It is of a dull grayish-yellow or brownish color that appears darker because of the numerous black punctures scattered over the surface. A smooth raised yellow line is noticeable running from the tylus nearly to the tip of the scutellum. It is common in fields of blue grass, timothy, etc. The range is from coast to coast in the northern states.

N. sulcifrons Stal somewhat replaces *undata* in the southern states. It can be distinguished by having a notch present between the postero-lateral angles of the juga and the eye, and the juga are also more strongly swollen. It is a smaller darker species without a light median stripe on the scutellum.

Figure 80

Families of Hemiptera   53

N. cavifrons Stal (fig. 80) is a dark western species occurring at most only slightly east of the Mississippi. It is readily separable from *sulcifrons*, which it closely resembles, by having the front of the head deeply concave, giving it a "scooped-out" appearance (the head in *sulcifrons* is shallowly concave). Two additional species occur in this country.

23a Juga not extending beyond and meeting before tip of tylus .... *Coenus*

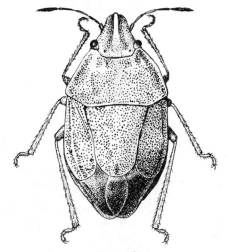

Figure 81

Figure 81 *Coenus delius* (Say)

This is a small (4.5-6 mm) dull straw yellow stink bug with strongly contrasting black punctures. The body is considerably broader across the middle of the front wings than across the pronotum. It is usually found on various grasses. *C. delius* is common in the Northeast but scarce elsewhere, and the southern limits of the range are poorly defined. One additional species is known from Arkansas and Oklahoma.

24 Dorsal surface thickly clothed with elongate upright silky or woolly hairs 25

24a Dorsal surface nearly glabrous, never conspicuously clothed with silky or woolly hairs 26

25 Lateral abdominal margins evenly rounded; labium extending to or beyond hind coxae .... *Trichopepla*

Trichopepla semivittata (Say)

This dull yellowish-brown stink bug is readily recognizable by the numerous upstanding hairs on the dorsal surface. The head is elongate with a dark stripe on either side of the tylus and another along the lateral margin of each jugum. It feeds on various species of umbelliferous plants. The distribution is over most of the United States east of the Rocky Mountains.

T. californica V. D. has a relatively shorter head (not longer than width of head across the eyes) and a somewhat shorter labium that only reaches the anterior margin of the hind coxae, whereas in *semivittata* the labium reaches or exceeds the posterior margin of the hind coxae. This species is confined to the northwestern states and British Columbia.

There are 6 additional North American species, most of them confined to the western states.

25a Postero-lateral corner of each abdominal segment produced out into a prominent tooth; labium short, not attaining mesocoxae .... *Prionosoma*

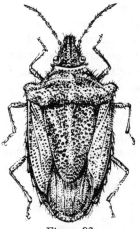

Figure 82

Figure 82 *Prionosoma podopioides* Uhler

This is a medium-sized brown species with a dense covering of rather woolly appearing hairs on the dorsal surface. Each humeral angle of the pronotum is produced laterally into a broad subacute tooth. The toothed lateral abdominal margin is distinctive. The distribution is over much of the western United States east to Missouri and Iowa.

26 Metathoracic scent gland area produced into an elongate tapering channel ....
.... *Holcostethus*

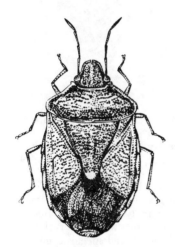

Figure 83

Figure 83   *Holcostethus limbolarius* Stål

This dark grayish-yellow species is thickly covered with dark punctures. The lateral pronotal margins, basal third of the lateral corial margins and the tip of the scutellum are strongly contrasting ivory white. It is a small stink bug, 7-8 mm in length. The species is commonly found in weedy fields over nearly the entire country, but in the western states is less common and is largely replaced by *H. abbreviatus*.

H. abbreviatus Uhler is closely related to *limbolarius* but the abdominal connexivum has an alternating pattern of dark and light patches, whereas in *limbolarius* the connexivum is chiefly black with a strongly contrasting uniformly pale yellow lateral margin. This is a common species from the Pacific coast states eastward to the Great Plains.

Five additional species occur in North America.

26a Metathoracic scent gland orifice ending in a short, rounded, ear-like lobe 27

27 Hind tibiae above with a distinct groove or channel present for entire length ....
.... *Euschistus*

This is a genus of medium large dull brown stink bugs. Several of the species are among the most common members of the family in most parts of the country and, to many people, represent THE stink bug. There are at least 18 North American species, some of which are economically injurious to fruits and truck crops. The species most likely to be collected may be separated by the following key.

a Ventral abdominal segments lacking black spots at anterior lateral angles. Fig. 84 .... *E. variolarius* (P. B.)

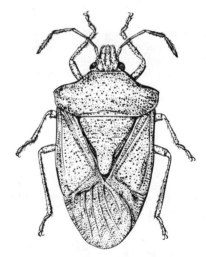

Figure 84

Males of this species may at once be recognized by the presence of a large black spot on the genital capsule. This is the "brown stink bug" of economic literature and is often destructive to truck crops and fruit. It is the most common species of stink bug in most of the northern states, but it is scarce in the South.

Families of Hemiptera   55

aa Abdominal segments below with small distinct black spots on anterior lateral angles .... b

b Abdomen below with a series of black spots down the midline .... *E. tristigmus* (Say)

This is the smallest of the common species of the genus. The number of black abdominal spots is variable, but there is always at least one anterior to the genital capsule. The range is over much of the United States east of the Rocky Mountains.

bb Abdomen below lacking a median series of black spots .... *E. servus* (Say)....c

c Juga considerably surpassing tip of tylus .... *E. servus euschistoides* (Voll.)

This subspecies occurs over most of the northern United States and southern Canada.

cc Juga not or only very slightly exceeding tip of tylus .... *E. servus servus* (Say)

This is the most common species of *Euschistus* in the southern United States. Intergrades between *servus servus* and *servus euschistoides* are frequent where their ranges overlap. It sometimes becomes destructive to cotton plants.

27a Hind tibiae rounded on all sides, lacking a deep groove or channel on upper surface 28

28 Humeral angles of pronotum with sharp prominent spines .... *Oebalus*

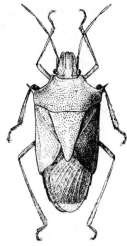

Figure 85

Figure 85 *Oebalus pugnax* (Fabr.)

The "rice stink bug" is a relatively elongate and slender insect, straw yellow in color, 9-11 mm long and readily recognizable by the prominent humeral spines. It is an extremely destructive species to rice, wheat, sorghum and many grasses. It occurs commonly east of the Rocky Mountains, north to Minnesota and New York.

Two additional species occur in the extreme southern states.

28a Pronotal humeral angles rounded or at mostly bluntly produced .... *Mormidea*

Mormidea lugens (Fabr.)

This is a small stink bug (6-7 mm) that can easily be recognized by the dull bronzy coloration with strongly contrasting white margins on the scutellum. It usually is found on bluegrass and related grasses. It is a very common species everywhere east of the Rocky Mountains. Three or four additional species are known from the southern and southwestern states.

COREIDAE
The Squash Bug Family

The members of this family are usually rather elongate but moderately heavy-bodied insects. Most species are predominately dull brown or gray in color and are of large or medium size. The membrane of the front wing has numerous veins. Frequently the hind legs have intriguing expansions and in certain species some of the antennal segments are similarly modified. All are plant feeders and some cause considerable damage to cultivated crops.

There are 33 genera of Coreidae known from the United States, but many are found only in the extreme southern or southwestern states.

1   Third antennal segment strongly dilated and flattened. Figs. 86, 87 2

1a Third antennal segment either cylindrical or three-sided, never strongly dilated into a flattened plate 3

2   Very large insects 35-40 mm; first and fourth antennal segments about equal in length; hind femora armed below with numerous teeth ....

Figure 86

Figure 86 *Thasus acutangulus* Stål

This is our largest coreid and among the largest of N. American insects. It is a very beautiful dark brown species with the hemelytral veins a contrasting orange-yellow and the antennae variegated red and dark brown. The hind tibiae are somewhat dilated. The males have the hind femora greatly enlarged. It is found on mesquite in Arizona and New Mexico.

2a Much smaller insects, usually not exceeding 15 mm; first antennal segment very elongate, much longer than fourth; hind femora bearing only a single spine below near distal end .... *Chariesterus*

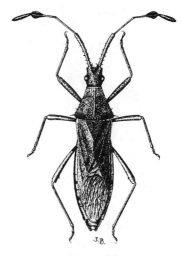

Figure 87

Figure 87 *Chariesterus antennator* (F.)

This is a slender-bodied species of dark brown color and covered with yellowish-brown or silvery hairs flattened against the body surface. The humeral angles of the pronotum are spined and the very long first antennal segment is somewhat three-sided. It occurs upon many flowering weeds especially *Euphorbia* spp., but the life history is poorly understood. The distribution is over the eastern and central states west to Colorado, but it becomes scarce and scattered north of the middle Atlantic states, Illinois and Iowa.

Three additional species are found in the southwestern states.

3   Hind tibiae strongly dilated and expanded, often having a "leaf-like" appearance. Figs. 88, 90 4

3a Hind tibiae sometimes slightly flattened, never strongly dilated and expanded 5

4   Tylus flattened from side to side (compressed) between juga and projecting upward as an acute projection .... *Acanthocephala*

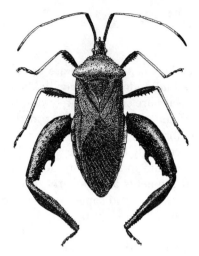

Figure 89

A. femorata (F.) (fig. 89) is an even larger species (25-28 mm) with the leaf-like expansions of the hind tibiae extending at least two-thirds of the way to the distal ends. The males of this species have the hind femora greatly enlarged and thickened, much more so than in the case of the females. It is a common species in the southern United States. Two or three additional species occur in the U. S.

4a Tylus compressed but extending outward in front of head, never curved upward as a spine or sharp projection .... *Leptoglossus*

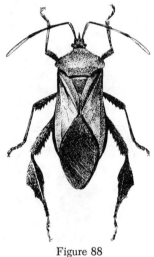

Figure 88

Figure 88   *Acanthocephala terminalis* (Dallas)

This is a large showy dark brown insect up to 22 mm in length. The color is often bright brown on the dorsal surface. The first three antennal segments are dark reddish-brown to chocolate-brown and the fourth is a strongly contrasting orange-yellow. The leaf-like expansions of the hind tibiae do not reach the distal ends of the tibiae. It occurs on many trees, shrubs and weeds. The eggs are of a beautiful golden-bronze color and oval in shape. *A. terminalis* is a common species over most of the eastern states and west to Colorado and Texas.

Figure 90

Figure 90 *Leptoglossus phyllopus* (L.)

This is a bright brown insect, 18-20 mm in length, with a striking white band running completely across the wing covers and with white markings on the leaf-like tibial expansions. It sometimes damages cotton, various fruits and truck crops. It is one of the most common coreids in the southeastern states and occurs in small numbers north to New York and west to Missouri.

L. oppositus (Say) is similar in size and color to *phyllopus* but is without a distinct white band across the wings (sometimes a trace of white markings is present). The lateral pronotal margins behind the humeral angles are smooth rather than having fine teeth or tubercles present. The distribution is from southern New England south to Florida, west to Minnesota and southwest to Arizona.

L. occidentalis Heidemann also lacks the distinct, complete transverse fascia across the fore wings. It differs from *oppositus* by having the hind tibial dilations lanceolate and without deep emarginations along the outer margin. This species breeds on conifers and is often destructive to the seed crop. It is a common species in the western mountain states from British Columbia to Arizona and extends eastward to Iowa and Kansas.

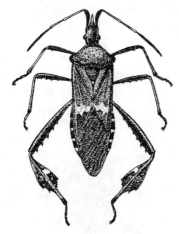

Figure 91

L. clypealis Heidemann (fig. 91) is readily distinguishable by a long spine that projects forward from the tip of the tylus. It occurs from Iowa, Missouri and western Texas to the west coast.

There are seven additional species in the United States, chiefly southern in distribution.

5   Second and third antennal segments three-sided .... *Chelinidea*

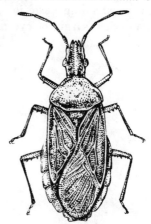

Figure 92

Figure 92 *Chelinidea vittiger* Uhler

This is a very distinctive light yellow species (11-15 mm) with pointed juga and peculiar three-sided antennae. The head is frequently dark chocolate-brown with a wide central yellow stripe. This species is very variable in color and several varieties have been formally described. The body is broader and more oval than is that of most coreids. It feeds on the prickly pear cactus (*Opuntia* spp.) and was introduced into Australia during the extensive biological control work done there on prickly pear. In this country *vittiger* is distributed throughout the southern states from Virginia and Florida to California in localities where its host plant occurs.

5a Second and third antennal segments not 3-sided, usually cylindrical 6

6   Antennal segments 1-3 clothed with a heavy conspicuous mat of elongate hairs. Fig. 94 7

6a First three antennal segments bearing only inconspicuous scattered hairs 8

Families of Hemiptera

7   Large reddish insects (16-21 mm), humeral angles of pronotum broadly rounded; bucculae extending posteriorly well beyond anterior margin of eye .... *Archimerus*

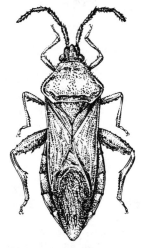

Figure 93

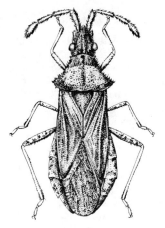

Figure 94

Figure 93   *Archimerus alternatus* (Say)

This species resembles *E. galeator* in color but tends to be slightly lighter and larger (16-21 mm). The antennae are heavy and very hairy as compared to *E. galeator*. The lateral margins of the connexiva dorsally have the anterior third yellow. The femora have bristly black hairs present. *A. alternatus* is generally an uncommon species although it is sometimes locally common especially on *Desmodium acuminatum* in fields and along woodland margins. It ranges from New Jersey south to Florida and west to Colorado.

Three additional species are found in the United States, two in Florida and one in the southwest.

7a   Relatively small (7-9 mm) grayish insects; humeral angles of pronotum produced as short acute spines (fig. 94); bucculae not extending posteriorly to reach anterior margin of eye ....*Coriomeris*

Figure 94   *Coriomeris humilis* (Uhler)

This is a relatively small species. It is grayish white above with numerous dark punctures and with a narrow white lateral pronotal margin. The range is chiefly west of the Mississippi, but it does occur rarely in the northeast. There is at least one additional western species known from Colorado.

8   Hind femora club-shaped, base very slender, becoming strongly thickened distally (fig. 95); hind tibiae with a thick short black spine at tip opposite point of attachment of tarsus .... *Merocoris*

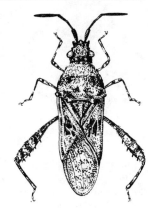

Figure 95

60   Families of Hemiptera

Figure 95 *Merocoris distinctus* Dallas

This is one of the smallest (7.5-9 mm) of the coreids found in this country. It has a short stubby grayish body covered with a thick matting of hair. The head is short and somewhat square. It is not especially common but usually taken on various weed flowers, especially goldenrod. One interesting record notes that many specimens were found on a dead chicken. It ranges over most of the United States west to the Great Plains. Two additional species are known but all are so closely related that their status needs clarification.

8a Hind femora sometimes curved or swollen but not club-shaped; hind tibiae lacking a spine that projects from tip; usually with a series a short acute comb-like teeth present 9

9 Hind coxae very close together at midline, separated from one another by less than width of a coxa .... *Ceraleptus*

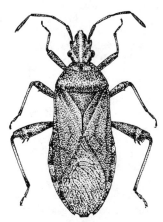

Figure 96

Figure 96 *Ceraleptus americanus* Stål

This is a medium-sized (9-11 mm) grayish-brown species with numerous dark punctures on the body from each of which arises a coarse scale-like hair. The humeral angles are evenly rounded. The head has a complete dull yellow central stripe. The range is a wide one in the southern states from Washington, D. C. south and west to Florida, Texas and California but records are very scattered.

Three additional species occur in the far west and Texas.

9a Metacoxae well separated mesally 10

10 Hind femora below bearing a series of large spines or teeth 11

10a Hind femora unarmed below, or at most with 2-3 small inconspicuous spines 12

11 Each antenniferous tubercle with a prominent spine present on outer side .... *Euthochtha*

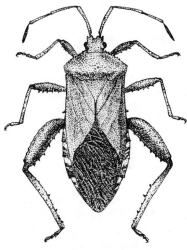

Figure 97

Figure 97 *Euthochtha galeator* (F.)

This is a moderate-sized (14.5-17 mm) dull brown species. It is stout-bodied with brownish antennae of which the fourth segment is usually darker. The humeral angles of the pronotum are rounded. *E. galeator* is frequently swept from wood margins, hedgerows and weedy fields where it feeds on a wide variety of plants. It is a common species throughout the entire eastern United States west to the Great Plains.

11a Antenniferous tubercles lacking a spine on outer side .... *Mozena*

Families of Hemiptera

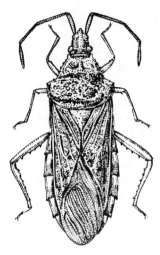

Figure 98

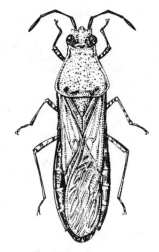

Figure 100

Figure 98 *Mozena obtusa* Uhler

This is a bright cinnamon-brown species with the scutellum and the anterior one-half of each abdominal connexivum and the abdominal venter yellow. It is a fairly common species on mesquite in the southwest. About 7 additional species occur in the southwest. *M. obesa* Mont. occurs from Kansas to Florida.

12 Metathoracic scent gland auricle with an anterior wrinkled raised area but this not forming a distinct rounded disc-like tubercle (fig. 99); with a raised ridge or platelike expansion of head immediately below antennal insertion ....
.... *Catorhintha*

Figure 99

Figure 100 *Catorhintha mendica* Stal

The color is a grayish-yellow with large dark punctures over the entire upper surface. The lateral margins of the connexiva dorsally have alternating orange and black bars. The distal 1/3 of the third antennal segment and the tip of the fourth are usually yellowish orange. The abdomen has numerous black spots. It breeds chiefly, if not exclusively, on wild four o'clock (*Mirabilis nyctaginea* [Michx.] MacM.) where it can often be taken in large numbers. It is fairly common in the middle west from Indiana to Colorado and south to Arizona. Careful study indicates that *mendica* has been following the eastward spread of its host plant along railroad embankments and it will be of interest to ascertain if this eastward movement continues in the future.

12a Metathoracic scent gland auricle with an anterior rounded disc-like tubercle (fig. 101); head lacking a raised ridge or platelike expansion immediately below antennal insertion .... *Anasa*

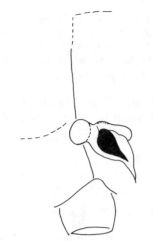

Figure 101

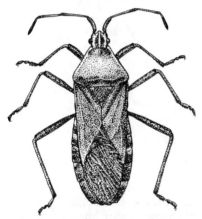

Figure 102

Figure 102 *Anasa tristis* (DeGeer)

This is the infamous "squash bug," one of the more serious insect pests that attack squash, pumpkin, cucumber and other members of the cucurbit family. It is a dull brownish-yellow insect (13-18 mm). The body is covered with large dark punctures. The head is dark brown with a yellow median and lateral pair of yellow stripes. The lateral margins of the connexiva dorsally have at least parts of the front half of each segment yellow. The head has a small tubercle behind the base of each antenna. It is distributed over most of North America.

A. armigera (Say) lacks a median yellow head stripe and has a long sharp spine at the base of each antennae that is nearly 1/3 the length of the first antennal segment. The fourth antennal segment is yellow or yellowish brown. The humeral angles of the pronotum are more prominently produced than in *A. tristis*. It usually lives on wild cucumber but occasionally injures cultivated cucurbits. The known distribution is from New England west and south to Iowa, Oklahoma and Texas.

A. repetita Heidemann is similar in size and appearance to *armigera* and like the latter has the fourth antennal segment a strongly contrasting orange-yellow, but *repetita* is more yellowish in color and lacks spines or tubercles at the antennal bases. It breeds on wild cucumber. It is a scarce species known from Massachusetts south to Virginia and west to Iowa, Missouri and Oklahoma. Three additional species occur in the United States.

ALYDIDAE
The Broad Headed Bugs

This family is closely related to the Coreidae. Most species have elongate slender bodies and broad prominent heads. They are all plant feeders and are usually found on the flowers and foliage of various weeds, although some appear to be restricted to plants of the grass family. There are scattered reports of carrion feeding. Many of the species particularly as nymphs closely resemble ants. The younger stages are so strikingly mimetic that when they are running on the ground in association with ants they "fool" even the experienced collector. Adults of certain species appear to be wasp mimics.

Twelve genera are known north of Mexico.

1   Hind femora very slender, without ventral teeth; juga extending well beyond apex of tylus and meeting or nearly meeting in front of it. Figs. 104, 105 2

1a Hind femora with one or two prominent rows of spines below; juga not extending forward beyond apex of tylus 3

2   Fourth segment of labium at least one and one-half times as long as the very short third segment; apex of juga forked as seen from side. Fig. 103 .... *Protenor*

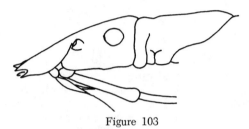

Figure 103

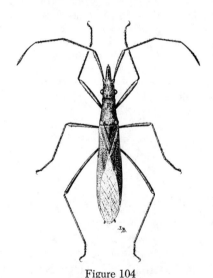

Figure 104

Figure 104 *Protenor belfragei* Haglund

This is a handsome, uniformly bright brown, elongate, parallel-sided, slender insect (12-15 mm.) with the antennae and often the legs and beak yellowish red. The dorsal surface is covered with dark punctures. It occurs in damp meadows on sedges and grasses. The range is northern west to the Rocky Mountains and in the east occurring south to Maryland. One additional species is found in the southern states.

2a Third and fourth segments of labium about equal in length; apex of juga not forked in lateral view .... *Leptocorisa*

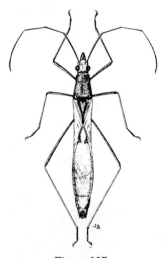

Figure 105

Figure 105 *Leptocorisa tipuloides* (De Geer)

This is an extremely slender elongate insect (14-16 mm) greenish yellow in life (fading to dull yellow) with greenish or reddish, yellow antennae. It is found on grasses and rice in many areas. In the Orient some members of this genus cause very serious injury to cultivated rice. It is restricted in distribution, in this country, to the extreme southern states from Florida to Texas. An additional species (*filiformis*) occurs in the southeast.

3   Body rather short and stout, usually not over 8 mm in length; scent gland opening absent or very small and without a distinct canal .... **Stachyocnemus**

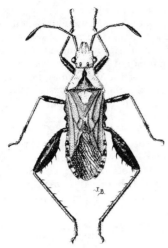

Figure 106

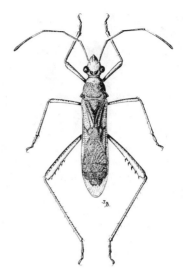

Figure 107

Figure 106 *Stachyocnemus apicalis* (Dallas)

This is a small species (up to 8 mm) with the color varying from grayish yellow to dark brown to black, frequently with reddish markings. Rather erect hairs are scattered over the body. The head is strongly bent downward. It occurs mainly in sandy regions and is a scarce species except in local situations. It is widely distributed (but very local) over most of the country but its western limits are poorly understood.

3a   Body elongate, rather slender, usually 10 mm or more in length; a distinct scent gland opening and auricle present in front of the hind coxae 4

4   Basal one-third to one-half of fourth antennal segment white, strongly contrasting with dark distal portion of segment .... *Megalotomus*

Figure 107 *Megalotomus quinquespinosus* (Say)

This is a dull reddish-brown insect (14-16 mm), sometimes varying to dull yellow. The proximal third of the fourth antennal segment is yellow and contrasts strikingly with the dark outer two-thirds. The humeral pronotal angles are angulate or somewhat pointed. It is usually found along the margins of deciduous woods and has been reported on *Baptisia, Ceanothus, Oxytropis, Lupinus, Rhus,* red clover, soybeans and *Desmodium*. The distribution is from coast to coast but it is not yet definitely known south of North Carolina in the East.

4a   Fourth antennal segment unicolorous throughout .... *Alydus*

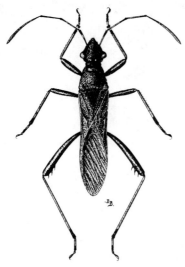

Figure 108

Figure 108 *Alydus eurinus* (Say)

This is a rather large (11-15 mm) black or very dark brown species. The lateral area of the abdomen has a small yellow spot near the front angle of each segment. The membrane is usually a solid dark color and the lateral pronotal margins are the same color as the rest of the pronotum. There are at least three records of *eurinus* on carrion (toad, cow). The relation to carrion would be an interesting problem for investigation to see if the insects obtain nourishment or only moisture from such a source. The nymphs of members of this genus are all striking ant mimics. Adults are frequently abundant on various weed flowers, and several species of the genus may be taken breeding together on *Amorpha canescens*. It is found throughout the United States but is most common in the northern states.

A. pilosulus (H. S.) is a slightly smaller, pale brownish tan species with the lateral pronotal margins white and the humeral pronotal angles sharp and acute. It is often found associated with *A. eurinus*. The distribution is across the entire country but it is most common in the southern states.

A. conspersus (Montandon) is a brownish to nearly black species often colored very much like *eurinus* and similar in size. It is most easily recognized by the numerous brown spots scattered over the membrane of the front wing. It is a widespread species chiefly northern but reported from southern Canada, California and Washington east to Pennsylvania and New England and south to Tennessee and northern Texas.

Three or four additional species occur in the northwestern states, one of them, *A. calcaratus* (L.), being common in the Rocky Mountain states.

REFERENCE
Fracker, S. B. 1918. The Alydinae of the United States. Ann. Ent. Soc. Amer. 11:255-280.

RHOPALIDAE
The Scent-less Plant Bugs

This is a small family of moderate sized hemipterans which is closely related to the Coreidae. There are numerous veins in the wing membrane and closed cells in the corium. The antennae are placed above the midline of the head, with the first segment shorter than the head length. The pronotum has a transverse ridge. The suture between abdominal terga five and six curves anteriorly. The thoracic scent gland orifices are very small. The large tuberculate ocelli are also distinctive.

Most of our species are found in weedy fields, although a few, including the abundant and wide spread Box Elder Bug, are arboreal. All species are phytophagous. The smaller species closely resemble species of the lygaeid genus *Nysius*, and the beginning student should carefully orient the light to observe the number of membranal veins present.

1   Hind femora bearing a series of prominent spines 2

1a   Hind femora lacking prominent spines
     3

2   Margins of abdomen not extending laterally distinctly beyond lateral margins of wings .... *Harmostes*

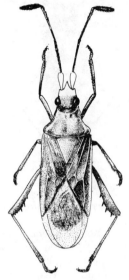

Figure 109

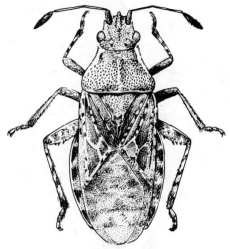

Figure 110

Figure 110 *Aufeius impressicollis* Stål

This insect resembles a small *Harmostes* (length 5-6 mm). It is grayish brown in color with a yellowish head. The abdominal connexivum dorsally is barred with alternate yellow and dark red-brown markings. An elongate black band is present behind each fore coxa. It is found in weedy fields throughout the western states east to Ohio.

Figure 109 *Harmostes reflexulus* (Say)

This is a moderately large species (7-9.5 mm) which is extremely variable in color from pale green to red-brown, often with brown spots scattered over the body surface. The lateral pronotal margins are strongly reflexed and without prominent teeth. It is a common species usually found on herbs in fields. The distribution is throughout North America.

H. serratus (F.) is a dark brown species with distinctly serrate lateral pronotal margins. It is found in the southern and southwestern states.

Six additional species occur in the southern and southwestern states.

Xenogenus extensum Dist. is a superficially similar species found in the southwest which may be distinguished from the species of *Harmostes* by the lack of produced processes at the antero-lateral pronotal angles.

2a Lateral margins of abdomen extended prominently beyond lateral margins of wings .... *Aufeius*

3   Lateral pronotal margins distinctly notched just behind anterior margin (figs. 112, 113); corium opaque; 11 mm or more in length 4

3a Lateral pronotal margins sometimes curved but never with a distinct notch behind anterior margin (fig. 111); corium at least in part semi-transparent (hyaline); usually at most little more than 9 mm in length 5

Figure 111

Families of Hemiptera   67

4   Bucculae not extending posteriorly more than half-way to base of head .... *Leptocoris*

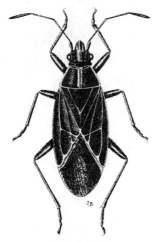

Figure 112

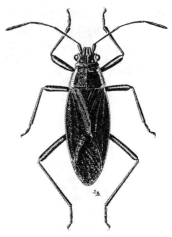

Figure 113

Figure 112   *Leptocoris trivittatus* (Say)

This is a large (11-14 mm) black insect with conspicuous red marking along the lateral pronotal margins, the apical and lateral corial margins, and as a stripe down the center of the pronotum.

This is the abundant and well-known "box elder bug" that occurs in great numbers on its host plant the box elder tree. In the fall these bugs often congregate in masses on the sunny sides of buildings and frequently become a nuisance by entering houses for hibernation. The box elder bug was originally a western species but has been steadily spreading eastward and is now widely distributed in the eastern states.

A second species occurs in California and Arizona.

4a Bucculae reaching base of head .... *Jadera*

Figure 113   *Jadera haematoloma* (H.-S.)

This is a beautiful black insect which has the eyes and broad lateral stripes on the head and pronotum a strongly contrasting bright red color (length 10-14 mm). Nothing seems to be known of its habits. The distribution is primarily southwestern in the United States, north to Iowa, Illinois and Colorado, but it also occurs in Florida and may be another example of a disjunct range. One or two additional North American species are known.

5   Impressed stripe in area of pronotal calli ending laterally in a distinct loop (fig. 111); posterior margin of metapleuron usually straight or slightly sinuate. Fig. 114 .... *Stictopleurus*

Stictopleurus punctiventris (Dallas)

This is a dull grayish-brown insect (6-12 mm) which has a broad black stripe on the first antennal segment and usually an impressed black spot between the eye and the ocellus. It occurs from coast to coast in the northern states.

S. viridicatus (Uhler) may be recognized by its smaller size (5-7.5 mm) and by its relatively light gray color. It is widespread in the western states east to Kansas and Nebraska.

Three additional species are recognized north of Mexico.

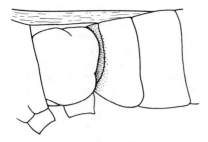

Figure 114

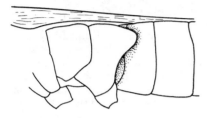

Figure 115

5a Transverse impressed stripe on pronotal calli not ending in a loop; posterior margins of metapleuron obliquely curved and angulate. Fig. 115 6

6 Pronotum anterior to transverse impressed stripe polished and impunctate, or with at most a few punctures on polished area .... *Liorhyssus*

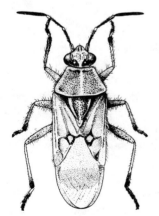

Figure 116

Figure 116 *Liorhyssus hyalinus* (F.)

This small species (5-6.5 mm) superficially resembles species of the lygaeid genus *Nysius*. The color is extremely variable from nearly black to reddish to pale yellow. The lateral pronotal margins are yellow and usually contrast with the darker color of the remainder of the pronotum. The tip of the corium is usually light reddish. The range is chiefly southern across the country. It is common in the South, less so in the North, but has been taken as far north as Massachusetts and Wyoming.

6a Pronotum anterior to transverse impressed stripe not smooth and polished, always with numerous coarse punctures present 7

7 Hind tibiae possessing a number of black annulate rings .... *Niesthrea*

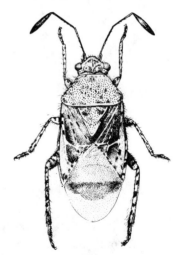

Figure 117

Figure 117 *Niesthrea louisianica* Sailer

This is a strikingly marked insect (6.5-8.5 mm) which varies in color from reddish to dark brown to grayish yellow and is thickly sprinkled with dark spots. It breeds on *Hibiscus* spp. It is scarce in the northern states. It ranges from New York south to Florida and west to Iowa, Texas and Arizona.

N. sidae (F.) is a very closely related species. It is somewhat smaller in size (usually between 5-6 mm) but can reliably be distinguished only by the male genitalia. The known range is from Florida and Georgia west to Texas.

7a   Hind tibiae often speckled or spotted but lacking a series of distinct annulate rings .... *Arhyssus*

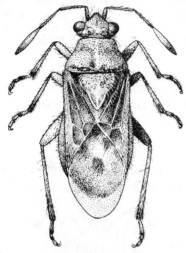

Figure 118

Figure 118   *Arhyssus lateralis* (Say)

This is a pale yellow to reddish-brown species, from 5-7 mm in length. The abdomen is usually yellowish. The species is extremely variable in color even in a single locality. Some specimens have the head and pronotum red, others are uniformly pale yellowish brown, and many color combinations are exhibited. It is frequently common on a variety of weedy plants in open fields. The range is almost throughout the United States and southern Canada.

A. nigristernum (Sign.) is generally a considerably darker reddish brown to dark brown species, distinguishable from *lateralis* by some femora (particularly the hind femora) having a "ringed" (or banded) appearance, and by having a subacute last (seventh) abdominal segment in the female (difficult to use in non-cleared specimens). In *lateralis* the apex of the last segment is almost rounded. It ranges from New England south to Florida and west to Louisiana and Kansas but is most common in the northeastern states.

Twelve additional species are recognized north of Mexico, the majority being widely distributed in the western states.

REFERENCES
Chopra, N. P. 1967. The higher classification of the family Rhopalidae (Hemiptera). Trans. R. Ent. Soc. Lond. 119:(12):363-399.
Chopra, N. P. 1968. A Revision of the Genus Arhyssus Stal. Ann. Ent. Soc. Amer. 61: 3:629-655.

LYGAEIDAE
The Seed Bugs

The Lygaeidae is one of the largest families of Hemiptera, with only the Miridae having a larger number of species in North America. The members of the family are exceedingly diverse in size, form and color, ranging from large gaudily colored black and orange-red species to obscure dull brown insects less than two mm in length. The great diversity of the family makes it difficult for the beginning student to recognize all species as belonging to a single family. Nevertheless, the Lygaeidae can be recognized by having four-segmented antennae that arise from the head at a point below an imaginary line drawn longitudinally through the eyes, by the presence of ocelli, and by having only five distinct veins in the membrane of the front wing.

The majority of lygaeids feed upon mature seeds of various plants, although several species are sap feeders and a few are predatory upon small insects.

1   General coloration red and black or orange and black, or if blackish then membrane with a large translucent white central spot connected to adjacent apical corial margin by a narrow band 2

1a Coloration variable, usually yellow, brown, reddish brown or black, but never highly variegated with strongly contrasting red or orange and black markings, if blackish then lacking a large white membrane spot connected by a band to apical corial margin 7

2 Scutellum raised and swollen, not strongly depressed on either side of elevated midline .... *Oncopeltus*

2a Scutellum with area adjacent to median ridge impressed, giving a "scooped out" appearance to scutellum on either side of midline, at least on posterior half 3

3 Pronotum with a broad red transverse stripe or three large red patches across posterior lobe, but posterior area of pronotum chiefly black .... *Lygaeus*

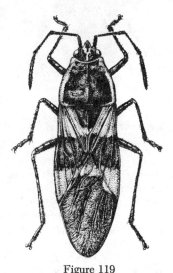

Figure 119

Figure 120

Figure 119 *Oncopeltus fasciatus* (Dallas)

This is a large (17-18 mm) orange and black insect known as the "large milkweed bug." The lateral margins of the pronotum, a Y-shaped head marking and two broad transverse bands across the wings are orange; the legs, antennae, wing membrane, central area of pronotum, a broad transverse band across the wings and all but the apex of the scutellum are dark brown to black. It is widely used in physiological work because of its hardiness and the ease with which it can be maintained on dry milkweed seeds in the laboratory. It can often be found in large numbers on the flowers and seed pods of milkweeds. It is widely distributed over most of the United States east of the Rocky Mountains, but is less common in the north. Six or seven additional species have been reported from the southern and southeastern states.

Figure 120 *Lygaeus kalmii* Stål

This species is known as the "smaller milkweed bug." It is a strikingly colored red and black species with a red spot at the base of the head. The membrane, clavus, and scutellum are black, and there is a large, irregular black patch near the middle of the lateral margin of the front wing. The red coloration of the wing forms a broad X-shaped area. In the eastern subspecies (spp. *angustomarginatus* Parshley) the wing membrane is either completely black or has a pair of small white spots. West of the 100th meridian specimens have a very large white membranal spot (spp. *kalmii* (Stål)). Like the "large milkweed bug" this species lives on milkweeds and is distributed over much of the United States.

L. turcicus (F.) resembles *kalmii* very closely in size and color. It is most easily distinguished by having a large Y-shaped red or yellow stripe on the head rather than merely a basal spot or short bar. It occurs in the northeastern and central states but the western limits of its distribution are poorly understood.

Four additional species are present in the western and southwestern states.

3a Pronotum either chiefly dark or if in large part red, then red coloration reaching hind margin mesally and not merely forming a broad sub-basal red band 4

4 Pronotum completely black .... *Melanopleurus*

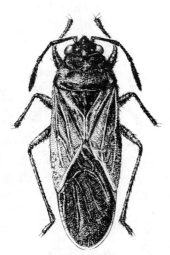

Figure 121

Figure 121 *Melanopleurus belfragei* Stål

The jet black head, pronotum, scutellum and wing membrane contrasting with the completely bright orange-red clavus and corium readily identify this beautiful species. It is 10 mm in length. *M. belfragei* is common in the southwestern states from Texas to California.

Three additional species occur in the southwest.

4a Pronotum not entirely black, always with pale or reddish areas at least along anterior margin or as spots at humeral angles and center of posterior margin 5

5 Scent gland auricle pale (yellow or red) .... *Ochrimnus*

Figure 122

Figure 122 *Ochrimnus lineoloides* Slater

This is a small (4.5-6.5 mm) orange and black species. The head is black with a tiny pale spot mesally near the base. The pronotum is orange with the anterior half having a large quadrate black area on either side of midline from which a dark ray extends posteriorly midway between the meson and the lateral margin to reach the posterior margin of the pronotum. The scutellum and wings are dark brown to nearly black with the lateral and apical corial margins and a central stripe on the posterior half of the scutellum orange-red. It is a common species from Virginia south into Florida and west to Texas.

O. mimulus Stål is similar in size and color but the large quadrate patches on the anterior pronotal lobe are red, the juga and an area near the eye are yellow, and the lateral and apical corial margins are usually pale yellow. The distribution is similar to that of *lineoloides*.

Five additional species occur in the west and southwest.

5a Scent gland auricle black 6

6 Membrane with a large white central spot connected by a narrow pale band to adjacent corial margin ...... *Lygaeospilus*

Figure 123

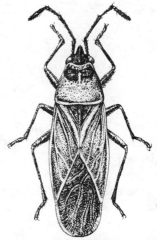

Figure 124

Figure 123 *Lygaeospilus tripunctatus* Dallas

This is a small insect (under 5 mm) chiefly black or dark brown in color with dull yellowish areas along the anterior pronotal margin and usually at the humeral angles. The membrane is black at the base, becoming reticulated gray and white distally with a prominent white discal spot. It is thickly clothed with grayish or silvery hairs. *L. tripunctatus* is a very widely distributed but usually scarce species that occurs from New England south to Florida and west to the Dakotas, Utah, Arizona and California.

Two additional closely related species occur in the southern and southwestern states.

6a Membrane lacking a white central spot, or if present, spot small and not connected to adjacent corial margin .... *Neacoryphus*

Figure 124 *Neacoryphus bicrucis* (Say)

This is a bright red insect (7-9.5 mm) with the head, legs, scutellum, wing membrane and large quadrate patches on the anterior pronotal lobe black. The bright red corium and posterior pronotal lobe make it easily recognizable. *N. bicrucis* has been reported from a number of different plants but its preferred food-plants remain poorly understood and needs further investigation. It is a common species throughout the southern United States but becomes scarce to rare north of New Jersey.

N. lateralis (Say) is similar in size and habitus but is of a dark gray-black color with the lateral corial margins, anterior margin, humeral areas and a caudo-mesal patch on the pronotum red or orange. It is frequently extremely abundant at lights. *N. lateralis* is a common species from the Great Plains westward.

Five additional species occur north of Mexico.

7 All sutures on ventral surface of abdomen reaching lateral margins of abdomen 8

7a Abdominal suture between third and fourth visible abdominal segments curving forward and becoming indistinct and obsolete before reaching lateral margin of abdomen. Figs. 16, 125 25

Figure 125

8   Eyes large and kidney-shaped (fig. 126), extending well posterior of antero-lateral angles of pronotum 9

8a Eyes variable in size and shape but never distinctly kidney-shaped, not extending considerably posterior of antero-lateral pronotal angles 10

9   Eyes stalked, entire area of eye posterior to level of ocellus located laterad of anterior pronotal angle. Fig. 126 .... *Hypogeocoris*

thickened and rounded with the corium enlarged and the membrane reduced which gives the insect a beetle-like appearance. It is widely distributed from New England south to Florida and west to Colorado and New Mexico.

Three additional species occur in the southern and western states.

9a Eyes very large but not prominently stalked, their posterior inner margins not located laterad of anterior pronotal angles. Figs. 127, 128 .... *Geocoris*

This large genus of "big-eyed" lygaeids contains approximately nineteen species north of Mexico. The most common and widespread species may be separated by the following key. These species are usually found running actively over the ground, generally in dry places. They are at least in part predatory and frequently are important in destroying economically destructive insects.

a   Vertex of head smooth and polished .... (Fig. 127) *G. punctipes* (Say)

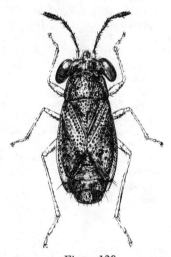

Figure 126

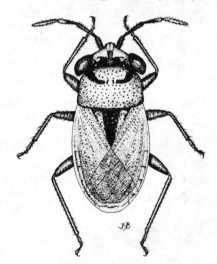

Figure 127

Figure 126 *Hypogeocoris piceus* (Say)

This is a beautiful black shining species (4 mm), usually with a strongly contrasting pale yellow or yellowish-red head and legs. It is usually found running on the ground in weedy open areas where it presumably feeds on other insects. The fully winged form is rare. Usually the front wings are

This is a grayish-to-yellowish species. It closely resembles *bullatus* in color including the pale stripes on the scutellum. In addition to the smooth shining head without granulose or rugulose surface *punctipes* can readily be recognized by the presence of a longitudinal groove or sulcus extending down the middle of the vertex.

The distribution is chiefly southern, extending northward to Maryland and Illinois with scattered records northward.

  aa Vertex of head, at least in part, rugulose or granulose .... b

  b Scutellum usually bicolored, usually slightly longer than wide .... (Fig. 128) *G. bullatus* (Say)

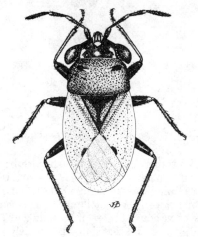

Figure 128

This is a yellowish-brown or griseus species generally with at least partially dark markings on the pronotal calli and on the head. The scutellum usually has a pale stripe on either side of the midline. Sometimes the scutellar marking is very much reduced. The range appears to include most of the United States.

    G. pallens Stal is a very closely related species but generally a lighter yellow in color and with an evenly convex scutellum (in *bullatus* the scutellum has a low median ridge). *G. pallens* is abundant in the western states east to the Great Plains.

  bb Scutellum usually unicolorous, nearly equilateral .... *G. uliginosus* (Say)

This is an extremely variable species from completely black shining with pale lateral corial margins to nearly pale tan in color. Several color varieties have been described the distribution of which should be analyzed as some at least appear to have geographic significance. It is often a very common species almost throughout the United States and southern Canada.

10 Membrane of fore wing with two closed cells present basally, outer cell the larger (adjust light angle carefully to see cells and veins) .... *Heterogaster*

Figure 129

Figure 129 *Heterogaster behrensi* Uhler

This dull brown and black insect (7 mm) is chiefly black below and on the head and anterior pronotal lobe. The legs and antennae are banded with orange or yellow and black. The eyes are rounded and set on short stalks. The dorsal surface is thickly covered with prominent silvery hairs. It is known only from the far western states. A second species is known from Louisiana and Texas.

10a Membrane of fore wing lacking two large closed basal cells 11

11 Fore femora greatly swollen and armed on ventral surface with numerous sharp spines 12

11a Fore femora either swollen or slender, usually with no more than two to three small ventral spines present, often without ventral spines (in *Plinthisus* several very small additional spines are present distally on fore femora; in this case pronotum is distinctly carinate along lateral margins and dorsal surface usually strongly shining) 14

12   Pronotum subtriangular, strongly tapering from posterior to anterior margins; lateral pronotal margins explanate (fig. 130), body very flattened; labium attaining or nearly attaining abdomen .... *Gastrodes*

Figure 130

Figure 130   *Gastrodes conicola* Usinger

This is a very broad and flattened dark brown species with an elongate-pointed head, and with the corial margins strongly expanded laterally. It is 6 mm or more in length. Members of the genus are unusual in that they live in the cones of coniferous trees where they feed on the seeds. *G. conicola* occurs in California.

Four additional species are found in the United States, three western and one reported from Ontario.

12a   Pronotum at most very moderately tapering anteriorly, lateral margins carinate or rounded, never explanate; body not strongly flattened; labium remote from abdomen 13

13   First antennal segment elongate, considerably exceeding apex of tylus; juga raised into a carinate ridge along lateral margins .... *Oedancala*

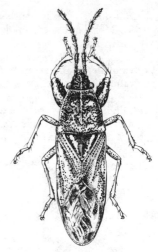

Figure 131

Figure 131   *Oedancala dorsalis* (Say)

This is a rather robust bright yellowish-brown species with a pale yellowish median stripe down the center of the pronotum and a pale calloused streak on either side of the scutellum. It feeds on the seed heads of various species of *Scirpus* and *Carex* in damp places. The distribution is widely over the northeastern and central states from New England south to northern Florida and west to Oklahoma and Colorado.

O. crassimana (F.) is found in the southern states in similar habitats and may be distinguished by having a considerably longer first antennal segment which is approximately as long as segments two and three combined, whereas in *dorsalis* the first segment is not more than two-thirds the length of segments two and three together.

An additional species has been taken in Texas and a fourth on the Florida Keys.

13a   First antennal segment short, at most only slightly surpassing tip of tylus; lateral margins of juga not raised and carinate .... *Phlegyas*

Figure 132

Figure 132 *Phlegyas abbreviatus* (Uhler)

This is a robust species (3-5 mm) of a reddish brown color. The legs are banded with black and white. It is usually taken in the brachypterous condition in grassy meadows where it is frequently abundant. The distribution is throughout much of the United States, but it is rare and local in the southern and western states where it is largely replaced by *P. annulicrus* Stal, a dull pale yellow species with a longer second antennal segment which is considerably more than twice the length of segment one, in contrast to *abbreviatus* where segment two is only twice the length of the first segment.

14   Spiracle present ventrally on seventh abdominal segment 15

14a Spiracle absent ventrally on seventh abdominal segment 18

15   Lateral surface of prothorax usually of a pruinose texture; spiracles of abdominal sterna 2-6 located dorsally 16

15a Lateral surface of prothorax non-pruinose in texture, smooth or granulose, punctate; all abdominal spiracles located ventrally 17

16   Apical corial margin straight; fore coxal cavities closed. Fig. 133 ...... *Ischnodemus*

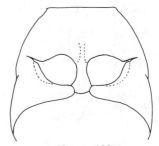

Figure 133

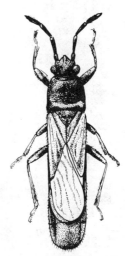

Figure 134

Figure 134 *Ischnodemus falicus* (Say)

This is an elongate, very slender, parallel-sided lygaeid with black head, pronotum and antennae and contrasting pale brown wings with dark veins in the membrane. It is a locally common species that often occurs in large colonies on *Spartina pectinata* in marshy areas throughout the eastern states west to the Dakotas and Texas, but records are lacking from most states south of North Carolina.

Ten additional species occur in North America, the majority being found in the southeastern and Great Plains areas.

16a Apical corial margin sinuately concave; fore coxal cavities open. Fig. 135 .... *Blissus*

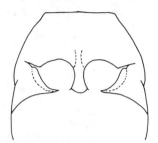

Figure 135

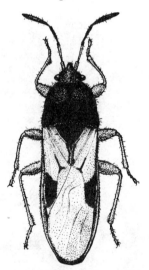

Figure 136

Figure 136 *Blissus leucopterus* (Say)

This is the infamous "Chinch Bug," one of the most destructive American insects which, during periods of abundance, has caused millions of dollars worth of damage to corn and small grains in the midwestern states. It is a small gray-black insect with white wings, and is distributed generally throughout the midwest.

B. insularis Barber is a very similar species of major importance as a pest of St. Augustine grass in Florida and the adjacent Gulf states.

B. leucopterus hirtus Montandon is an eastern subspecies of the chinch bug which is a serious pest of lawn grasses in the northeast.

A dozen additional species occur, most of them closely resembling *leucopterus* in appearance. One or another of the various species is found nearly everywhere in the United States and southern Canada.

17   Bucculae very short and rounded, not extending posteriorly as far as base of antenniferous tubercles; head smooth or with very small obscure punctures; usually brachypterous in which case width across anterior pronotal lobe greater than width across humeri .... *Plinthisus*

Plinthisus americanus (V. D.)

This is a small (3-3.5 mm) oblong, shining chestnut to reddish-brown species, usually with the anterior pronotal lobe convex and in brachypters broader than the posterior lobe. It is usually found in litter on the ground, and in contrast to most ground-living lygaeids is a forest species occurring in litter beneath hemlock and spruce fir. It is usually densely covered with upright hairs (easily abraded). It is known only from southeastern Canada, New England and New York, but is probably more widely distributed in the eastern mountains.

Four additional species are known from the western and southwestern states.

17a Bucculae elongate, extending nearly to base of head, far exceeding base of antenniferous tubercles; head always coarsely and deeply punctate; width across humeri of pronotum always greater than width across anterior lobe .... *Crophius*

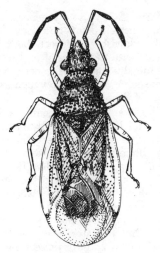

Figure 137

Figure 137 *Crophius disconotus* (Say)

This is a small (3-4 mm) species, rather elongate-oval in shape, with the head, pronotum and scutellum dark brown and the fore wing a strongly contrasting somewhat hyaline white. The fore wings are much wider than the abdomen and extend posteriorly beyond its apex. It occurs chiefly on goldenrod and *Pinus* spp. *C. disconotus* is a scarce species but is widely distributed from Canada and New England south to Alabama and west to Utah and California.

    C. bohemani Stål is a larger, more robust species with a conspicuous white anterior band on the pronotum. It is found in the extreme western United States.

    Members of this genus are similar in shape to species of the genus *Kleidocerys*.

    Seven additional species occur in the western states.

18   Clavus lacking numerous punctures 19

18a   Clavus with numerous distinct punctures present 22

19   Fore femora with a sharp spine below on distal third .... *Belonochilus*

Belonochilus numenius (Say)

This is an elongate dull yellow insect (5-7 mm) usually with a dull reddish-brown patch at the apex of the corium, small brown spots on the femora and a very elongate labium which reaches at least to the middle of the abdomen. It feeds chiefly on the seed heads of Sycamore. *B. numenius* is an uncommon species that is widely distributed over most of the United States east of the Great Plains and has also been reported from Arizona and California.

19a   Fore femora unarmed; head often not extending forward beyond distal end of antennal segment one 20

20   Lateral corial margins straight throughout entire length .... *Ortholomus*

Ortholomus scolopax (Say)

This is a dull grayish-brown insect 5-6 mm in length, with the membrane subhyaline and the apices of the coria usually dull reddish. It closely resembles species of *Nysius* in general appearance. The tylus somewhat exceeds the distal end of the first antennal segment. The dorsal surface is thickly clothed with short silvery hairs. This species frequently is very common on dry weedy fields almost throughout the United States and southern Canada.

    Four additional species occur in the southern and western states.

20a   Lateral corial margins expanded outward a short distance from base, thus not parallel-sided throughout entire length 21

21   Bucculae short, extending posteriorly at most only slightly beyond middle of eye, usually scarcely reaching eye .... *Xyonysius*

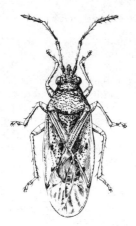

Figure 138

Figure 138 *Xyonysius californicus* (Stål)

This is a robust gray-yellow species (4.5-6 mm) with scattered brown and black markings on the wings and the pronotum. It is found in dry fields nearly throughout the United States.

21a Bucculae longer, reaching or almost reaching base of head, extending well beyond middle of eye .... *Nysius*

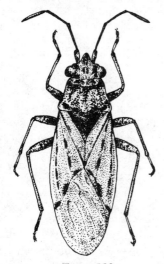

Figure 139

Figure 139 *Nysius niger* Baker

This is the "false chinch bug" and is one of the most common species of lygaeids in many parts of the United States. It is a very general feeder, attacking many garden plants and some fruits. It is a small, dull yellowish-gray insect about 3-4 mm in length, usually with two series of broken spots running longitudinally through the corium and with a glassy transparent membrane which extends beyond the end of the abdomen.

The taxonomy of the genus is extremely complicated and has only recently been carefully studied so that the ranges of the various species as given in the literature are untrustworthy, as they are frequently based on misidentification. Most records refer to this species under the name *Nysius ericae* (Schilling) which actually is found only in Europe.

Nine additional species are now thought to occur in the United States, many of them widespread and common. Most of these are very similar to one another in size and shape and can only be reliably distinguished by dissection and critical study of the genitalia.

22 Corium translucent, hyaline, lacking distinct punctures over most of translucent surface; head lacking a curved sulcus before each ocellus 23

22a Corium opaque, not translucent, having numerous punctures over nearly entire surface; a curved longitudinal sulcus present before each ocellus 24

23 Apex of scutellum upturned and bifid; corium strongly constricted laterally on basal half .... *Cymoninus*

Cymoninus notabilis (Distant)

This is a small (3-3.5 mm) slender delicate insect with a brown head and pronotum, glossy wings, a green (fading to yellow) abdomen, and with a coarsely densely punctured pronotum. It occurs on sedges in Florida, Georgia and the Gulf Coast states where it is often abundant.

23a Apex of scutellum pointed, not bifid; basal half of corium not strongly sinuately constricted .... *Kleidocerys*

Kleidocerys resedae (Panzer)

This is an ovoid glossy-winged reddish-brown insect with the wings greatly exceeding the abdomen both laterally and posteriorly. The apical corial margin has a series of three or four small but conspicuous dark spots. It is often abundant in colonies on a variety of host plants, most frequently on wild spiraea, cat-tails, birch and rhododendron where it feeds on the seeds. The distribution is northern across the United States and Canada including Alaska and Newfoundland.

Six additional species occur, some widely distributed and sympatric with *resedae*. Species of this genus are all very similar and their separation critical.

24   First antennal segment elongate, exceeding apex of tylus .... *Cymodema*

Cymodema breviceps (Stål)

This is an elongate-oval species (4-4.5 mm in length). It is of a light yellow color and closely resembles species of *Cymus*, from which it is easily distinguished by the relatively elongate first antennal segment. It feeds upon sedges in marshy places. The range is southern from New Jersey and Indiana southward into Florida and at least to Texas. Much of the previous literature treats this species under the name *Cymus virescens* F.

24a   First antennal segment shorter, not exceeding apex of tylus .... *Cymus*

This genus contains nine North American small brown heavily punctate species that closely resemble the seeds of sedges and rushes, in the heads of which they are concealed. The following key will separate the more common species.

    a   Tylus greatly produced beyond anterior end of bucculae; second segment of labium only slightly exceeding anterior margin of prosternum ....
         .... (Fig. 140) *coriacipennis* (Stål)

(Distributed in the far western states east to Utah, Idaho and New Mexico.)

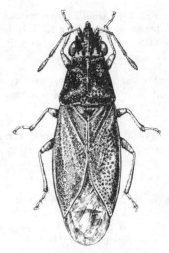

Figure 140

    aa   Tylus at most very slightly produced forward of bucculae; second segment of labium considerably exceeding anterior prosternal margin, reaching or nearly reaching fore coxae .... b

    b   Second antennal segment about 1/3 shorter than segment three ....
         .... (Fig. 141) *angustatus* Stål

(The distribution is nearly throughout the country east of the Rocky Mountains.)

Figure 141

    bb   Second and third antennal segments nearly equal in length .... c c Size very small, not over 3.5 mm; antennal segment two slightly longer than three; scutellar carina usually indistinct .... *discors* Horvath

(This is a northern transition zone species. It is common in southern Canada and the northern states east of the Rocky Mountains.)

cc Second and third antennal segments subequal in length; size larger, 4 mm or greater; scutellar carina usually distinct .... *luridus* Stål

(The distribution is chiefly northern from Nova Scotia and New England west to British Columbia and California. It is known from as far south as Georgia, Arkansas and New Mexico.)

25 Anterior pronotal lobe with lateral margins strongly and evenly rounded (fig. 142); with at most a very obsolete indication of a carina present (i.e. *Kolenetrus*) 26

25a Anterior pronotal lobe either carinate or explanate, never evenly and smoothly rounded. Figs. 159, 160 41

26 Head produced into a long slender "neck" (fig. 142); head one and one-half times length of pronotum ...... *Myodocha*

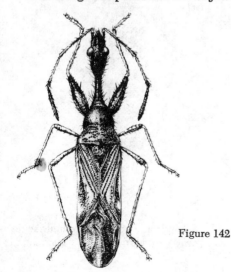

Figure 142

Figure 142 *Myodocha serripes* Olivier

This is a bizarre elongate, slender species with a shining black head and a very elongate neck. The pronotum is uniformly dull gray, and the wings dark brown with pale lateral margins. The legs are pale yellow with dark distal bands on the femora. It is found on the ground in relatively "new" habitats rich in forbes one to two feet high. It disperses in the fall to woodland margins where it often hibernates in loose associations under bark or in litter. It is occasionally injurious to strawberries. *M. serripes* is a common species from southern Canada and New England south to Florida and west to Colorado and New Mexico.

A closely related species occurs in Florida in similar habitats.

26a Head with at most a short "neck" present (fig. 148); head usually shorter than pronotum, if longer then head much less one and one-half times pronotal length 27

27 Head, pronotum and scutellum uniformly black (in *Cnemodus*, macropters sometimes have these areas dark reddish brown) 28

27a Head, pronotum and scutellum not uniformly black 29

28 Anterior margin of pronotum with a distinct collar .... *Cnemodus*

28a Anterior margin of pronotum lacking a distinct collar .... *Kolenetrus*

Figure 143 *Cnemodus mavortius* (Say)

This is a large (9 mm) elongate gun-metal black subshining to shining species with elongate strongly contrasting yellowish-brown legs. It has a very elongate anterior pronotal lobe, usually three times the length of the posterior lobe. It runs rapidly on the ground in dry areas with bare gravelly soil, particularly where *Andropogon scoparius* is abundant. This species is usually found in the brachypterous condition in which case the ocelli are lacking. It is uncommon but widepread, oc-

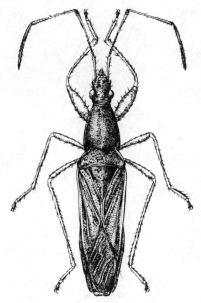

Figure 143

Figure 144

curring from New England south to Florida and west to Texas and Kansas.

Two additional species are recognized from the southern and central states.

Figure 144 *Kolenetrus plenus* (Distant)

This is a small (3-3.5 mm), black, shining species with large patches on the front wings and the antennae reddish brown. Males generally are orangish brown on legs and antennae. It is found on the ground in dry but cool old field ecotones. The distribution is disjunct in mountainous areas. It is known from southeastern Canada and northern New England and New York, British Columbia, Montana, Arizona and North Carolina.

29 Anterior pronotal lobe with a distinct complete collar (fig. 152), collar with a complete incised line along posterior margin; if collar relatively indistinct then length over 4.5 mm 30

29a Anterior pronotal lobe lacking a distinct complete collar, but sometimes with an impressed series of punctures in this region; if a vague collar present then size under 4 mm, usually less than 3.5 mm 38

30 Anterior pronotal lobe strongly shining, usually reddish tan in color, but if anterior lobe black shining then a prominent transverse white patch present on distal fourth of corium 31

30a Pronotum usually of dull texture or if subshining then anterior pronotal lobe black but in this case lacking a prominent white patch on distal fourth of corium 33

31 First antennal segment short, only slightly surpassing apex of tylus, always exceeding tylus but by much less than one-half length of first antennal segment .... *Sphaerobius*

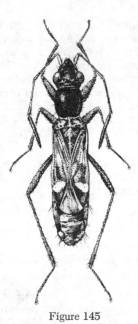

Figure 145

Figure 145 *Sphaerobius insignis* Uhler

This is a handsome ant mimic with a narrow swollen thorax, ant-like head, and with the basal half of the corium and a triangular patch near the corial apex whitish. The species is usually brachypterous and exhibits color polymorphism. Usually nearly half of the individuals in an area are dark brown to nearly black, the other half bright tan. They resemble large active ants and move in a very ant-like fashion. They are usually found in very dry hot barren habitats among sparse clumps of *Andropogon* at roadsides and railroad rights-of-way. This is a northern local species from the east coast to the Sierra Mountains.

A second species is known only from New Jersey.

31a First antennal segment elongate, surpassing apex of tylus by half or more than half length of segment 32

32 Apical fourth of corium always with a large white patch present .... *Pseudopamera*

Figure 146

Figure 146 *Pseudopamera aurivilliana* Distant

This is a large, striking, reddish-brown species (7.5-9 mm in length). The fourth antennal segment has a broad pale basal band. This is a common species in Texas and the southwest. Five additional species occur in the western and southwestern states.

32a Apical fourth of corium unicolorous with general corial coloration, never with a conspicuous white patch present .... *Pseudocnemodus*

Figure 147

84   Families of Hemiptera

Figure 147 *Pseudocnemodus canadensis* (Provancher)

This is an elongate slender reddish-brown shining species with pale lateral corial margins and largely yellow legs. The distal portions of the femora are usually a contrasting red-brown. It usually occurs in the brachypterous condition. The habitat is chiefly in dry overdrained grassy areas, particularly in edge habitats between woodlands and old fields. It is a northern species distributed from southern Canada and New England west to Kansas, Nebraska, the Dakotas, and British Columbia.

33 Head with a distinct but short "neck"; distance from posterior margin of eye to base of head over twice that from anterior eye margin to base of antenna .... *Heraeus*

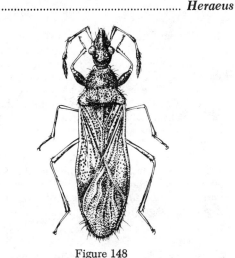

Figure 148

Figure 148 *Heraeus plebejus* Stål

This is a slender yellowish to reddish-brown species with wings variegated with yellowish and brown. It is 4.5-5.5 mm in length. This species is found most frequently in cool moist soils between fields and marshes or woodlands. It migrates for hibernation to woodland margins where it is frequently found with *Myodocha serripes*. This is a common widespread species over the eastern states west to the Great Plains and south to Florida and Arizona.

Four additional species occur in the southern and southwestern states.

33a Head without a prominent distinct "neck" present; distance from posterior margin of eye to base of head subequal to or less than distance from anterior margin of eye to base of antenna 34

34 Third and fourth abdominal sterna with a distinct curved lunate stridulatory area present (at low power appearing as a smooth shining curved area). Fig. 125 .... *Ligyrocoris*

Ligyrocoris diffusus Uhler

This is an elongate pale reddish-brown insect (5-7 mm in length). The posterior pronotal lobe usually has four obscure dark stripes present. The head and the anterior pronotal lobe are black or gray, and the corium has a pale spot near the inner apical angle. This is a very abundant and widespread species that is most common in open disturbed habitats such as old fields, particularly where numerous composites are present. It is a ground-living species but often climbs plants to feed on the seeds. It overwinters in the egg stage. It is widely distributed across nearly the entire country but is absent in the southeast.

L. sylvestris (L.) is a closely related species which is distinguishable by nearly lacking upright hairs on the pronotum and fore femora (conspicuously present in *diffusus*), and with the head black and the corium dark distally. *L. sylvestris* is usually slightly brachypterous, the fore wing not attaining the apex of the abdomen. It is a boreal species found only in the northern states and Canada or at relatively high altitudes further south.

Eleven additional species occur north of Mexico, several of which are widespread.

34a No stridulatory lunate area present on abdominal sterna three and four 35

35 Pronotal lobes separated by a shallow obtuse constriction. Figs. 149, 150 36

35a Pronotal lobes separated by a deep acute incision (fig. 152) (sometimes not entirely across midline) 37

36 First segment of hind tarsus three times length of segments two and three combined .... *Zeridoneus*

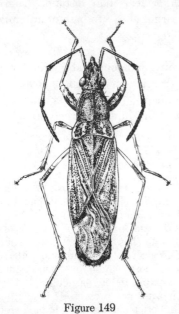

Figure 149

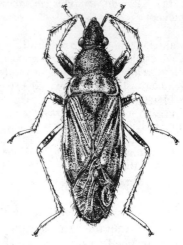

Figure 150

Figure 150 *Perigenes constrictus* (Say)

This is a moderately large robust species (7.5-9 mm) with the head, anterior pronotal lobe and the scutellum dull black, and the posterior pronotal lobe and the fore wings variegated with dark reddish brown. The body is thickly clothed with yellowish hairs. The distribution is throughout the northern and central states. It occurs in temporary habitats, even in places such as vacant lots and along roadsides.

P. similis Barber is a closely related species that largely replaces *constrictus* in the southern states. It is a somewhat smaller species with a shorter second antennal segment which is not longer than segment four.

Figure 149 *Zeridoneus costalis* (V. D.)

This is an elongate slender species (6-8 mm) that resembles species of *Ligyrocoris* in general appearance but lacks the abdominal stridulatory vitta and has a less deeply incised, transverse pronotal impression. It is found in open field habitats in early succession stages. This is a northern temperate species distributed in the northern states and southern Canada south to Iowa, northern Illinois and southern New England. Two additional species occur in the southwestern states.

36a First segment of hind tarsus little more than twice length of segments two and three combined .... *Perigenes*

37 Over 5.5 mm in length; corium nearly uniformly pale yellowish brown, at most with an obscure dark stripe running anteriorly from middle of apical corial margin .... *Paromius*

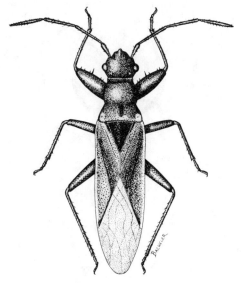

Figure 151

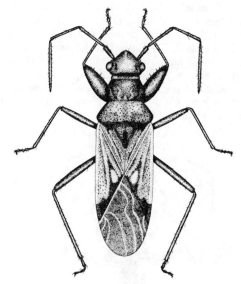

Figure 152

Figure 151 *Paromius longulus* (Dallas)

This is an elongate slender pale tan species with the head strongly exserted. It is often abundant on the ground as well as on plants, and sometimes is destructive to strawberries. *P. longulus* is a southern species ranging from North Carolina to Texas and is often one of the most abundant lygaeids in the southern states.

37a Usually not over 5 mm in length, if over 5 mm then corium with contrasting dark markings present .... *Pachybrachius*

Figure 152 *Pachybrachius vincta* (Say)

This is a very small (2.5-3.5 mm) lygaeid, dark gray-black in color, with pale wings bearing strongly contrasting black areas along the apical corial margins, and with a conspicuous oval white spot near the inner apical angle. This species is often destructive to strawberries. It is abundant from North Carolina south and west to Texas.

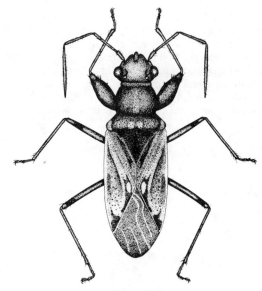

Figure 153

Families of Hemiptera   87

P. basalis (Dallas) (fig. 153) is a small (4 mm) species of a dull gray-black coloration with the hind lobe of the pronotum chiefly pale and finely punctate. The apical corial margin is pale with a small apical dark area. The lateral corial margin lacks a dark bar. This species is widely distributed over the northern and central states west at least to the Great Plains and south to Florida and probably Texas.

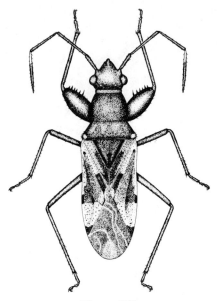

Figure 155

P. albocinctus Barber (fig. 155) is very similar to *bilobatus* but has the basal one-third to one-half of the fourth antennal segment white. The distribution is similar to that of *bilobatus*. Three additional species occur north of Mexico.

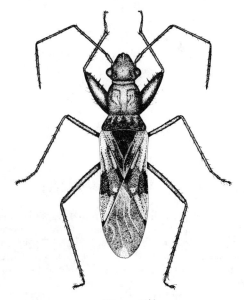

Figure 154

P. bilobatus (Say) (fig. 154) is a larger more elongate species (4.75-5.5 mm) and is distinguishable by having a dark spot along the lateral corial margin behind the middle, another spot at the corial apex and by having a completely dark fourth antennal segment. It is one of the most abundant ground-living lygaeids in the southeastern states, and it occasionally occurs northward.

38 Third and fourth antennal segments much enlarged and considerably broader than segment two. Figs. 156, 157 39

38a Third and fourth antennal segments either slender or stout, but third segment never strikingly broader than segment two. Fig. 158 40

39 Anterior pronotal lobe black or dark brown, strongly contrasting with pale posterior lobe, subshining; first two antennal segments usually yellow, segments three and four dark reddish or black; median scutellar carina pale ....
.... *Ptochiomera*

Figure 156

Figure 157

Figure 156 *Ptochiomera nodosa* Say

This is a handsome little species (2.5-3.5 mm) that has greatly enlarged third and fourth antennal segments which are in large part dark red or black, and strongly contrasting with the pale first and second segments. The head and anterior pronotal lobe are dark red-brown or black and the posterior pronotal lobe and the wings a contrasting pale yellow. It is often common on the ground in open fields, along roadsides, and in parking lots—essentially open dry habitats. *P. nodosa* is common in the southern states, and its range extends north to southern New England, Iowa and Illinois and west to Nebraska.

39a   Entire pronotum dull reddish brown, anterior and posterior lobes not strongly contrasting; all antennal segments dark reddish brown, scutellar carina not pale .... *Sisamnes*

Figure 157 *Sisamnes clavigerus* (Uhler)

This is another small (2.5-3 mm) short legged oblong-oval species with strikingly enlarged third and fourth antennal segments. It is nearly uniformly dark reddish brown in color, and coarsely punctate over the entire surface which also is clothed with short, broad, scale-like hairs. It is a scarce usually brachypterous species found on the ground in dry, sandy, warm fields. It ranges widely from New England south to North Carolina and west to Idaho and New Mexico. A second species is found in Florida and the southwest.

40   Antennae stout and thick, bearing prominent hairs at right angles to length of segments and as long as or longer than diameter of segment; males lacking a large preapical tibial tooth on fore tibiae .... *Carpilis*

Carpilis consimilis Barber

This is a tiny oval-oblong species (2-3.5 mm) with the head, anterior pronotal lobe and the scutellum a dark reddish brown, and the posterior pronotal lobe, the clavus, and the corium a strongly contrasting pale yellow. It is found on dry overdrained moraine slopes and on mountain balds, usually on cool north-facing slopes. The known distribution is northeastern from Quebec to New Jersey. Two additional species are known from Florida, Texas and New Mexico.

40a Antennae relatively slender, segments two-four lacking elongate prominent hairs as long as or longer than diameter of segment; males with a prominent preapical tooth on fore tibiae .... *Exptochiomera*

Figure 158

Figure 159

Figure 158 *Exptochiomera confusa* Barber

This is a small (3-4 mm) fuscous brown and yellowish oblong-oval species. The posterior pronotal lobe, the clavus and the corium are brownish yellow and thickly covered with dark brown punctures. It is a ground living species usually occurring on moist soil and common along the Gulf Coast from Florida to Texas. Seven additional species occur north of Mexico, all but one restricted to Florida or the southwestern states.

41 Dorsal surface strongly shining, pronotum and hemelytra with elongate upstanding hairs .... *Xestocoris*

Figure 159 *Xestocoris nitens* V. D.

This is a small (3-3.5 mm) shining dark brown species with somewhat paler reddish-brown legs and antennae. It is usually brachypterous and found on the ground among grasses on overdrained soils. The distribution is northern from New England west to Iowa and south in the mountains to North Carolina.

41a Dorsal surface either dull or shining, but if upstanding elongate hairs present on pronotum and hemelytra then surface is dull 42

42 Head, pronotum, and scutellum completely black .... *Atrazonotus*

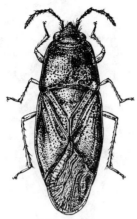

Figure 160

Figure 160 *Atrazonotus umbrosus* Distant

This is a medium-sized (5-7 mm) completely black species. It is found on the ground in a variety of habitats but usually in temporary fields with numerous weedy plants present. It is a somewhat uncommon species but is found nearly throughout the United States.

42a With at least posterior pronotal lobe having conspicuous pale markings 43

43 Head and central disc of anterior pronotal lobe black, strongly contrasting with pale markings on posterior pronotal lobe 44

43a Head and pronotum predominately pale or reddish brown, these areas never completely black 48

44 Lateral pronotal margins carinate (fig. 161), never expanded into a shelf-like explanate flange .... *Peritrechus*

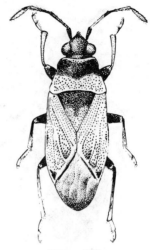

Figure 161

Figure 161 *Peritrechus fraternus* Uhler

This is an oblong-oval species (4-5 mm) with the posterior pronotal lobe, the clavus and the corium dull brownish yellow. It is most frequently found in wash litter communities and along new roadsides. *P. fraternus* is a widespread species across the entire country. It has been reported in the west from Mexico north to British Columbia. In the east it appears to have a northern distribution from Quebec south to New Jersey. It is common in the Midwest. At least three additional species occur north of Mexico.

44a Lateral pronotal margins expanded into a shelf-like explanate flange. Fig. 162 45

45 Explanate pronotal margins bearing a series of prominent bristle-like black hairs; scutellum with a pair of longitudinal pale stripes on posterior half, one on either side of midline .... *Sphragisticus*

Figure 162

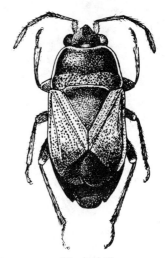

Figure 163

Figure 162 *Sphragisticus nebulosus* (Fallen)

This is an elongate oval species of moderate size (4-5.5 mm) with a dark head, pronotum and scutellum, with the lateral pronotal margins contrastingly pale, the corium grayish yellow mottled with dark brown, and with the membrane dull yellowish. It closely resembles species of *Peritrechus*. It is an abundant species in arable disturbed habitats such as the margins of cultivated fields. It is possibly an introduced European species. The distribution is northern from southern Canada south to New Jersey and west to British Columbia, Alaska and California.

Figure 163 *Trapezonotus arenarius* (L.)

This species is similar in size and shape to *Sphragisticus nebulosus* with the head, anterior pronotal lobe and scutellum black and the posterior pronotal lobe and the corium largely dull grayish yellow. The species occurs primarily in the brachypterous condition. It is found on upland, usually dry overdrained slopes, along roadsides and in alpine meadows. It is a northern species distributed from New England and Quebec west to British Columbia. Four additional species occur in the western states (the generic relationships of these species appear to be unsatisfactory).

45a Explanate pronotal margins lacking a series of black bristle-like hairs, if elongate hairs present these thin and pale colored; scutellum completely or nearly completely black 46

46   Pronotum lacking a deep transverse constriction, thus pronotum not separated into two distinct lobes; explanate lateral margins of pronotum of even width throughout .... *Trapezonotus*

46a Pronotum distinctly separated into two lobes by a strong transverse constriction; explanate pronotal areas distinctly broader at lateral ends of transverse pronotal constriction than along anterior pronotal lobe 47

47   Over 5.5 mm in length; numerous elongate, fine pale hairs present on dorsal surface .... *Eremocoris*

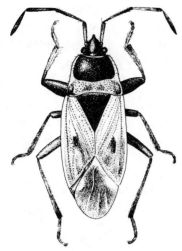

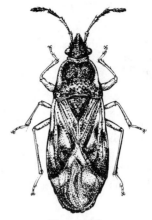

Figure 165

Figure 165 *Scolopostethus thomsoni* Reuter

Figure 164

Figure 164 *Eremocoris ferus* (Say)

This is a moderately elongate species with the head, anterior pronotal lobe and scutellum black. The posterior pronotal lobe is pale reddish brown with four somewhat irregular dark stripes present. The explanate edges of the pronotum are contrastingly pale. The corium is chiefly pale yellowish to brown with dark spots. The membrane is dark with a conspicuous light spot adjacent to each corial apex. It is a woodland species found in litter below hemlock, spruce and birch. *E. ferus* is apparently widely distributed but it seems probable that two species are involved, one southern and the other northern, and the distribution of *ferus* is uncertain pending further study. Ten additional species occur, chiefly in the western states.

This species has much the appearance and coloration of a small *Eremocoris*. It is found in woodlands throughout the northern states and southern Canada. However, the situation is complex with several different ecological habitats known such as *Carex* clumps, and further study is needed as a complex of closely related species may be present. Four additional species occur north of Mexico.

47a Not over 4 mm in length; with only short inconspicuous hairs present on dorsal surface .... *Scolopostethus*

48 Anterior pronotal margin with a distinct complete collar present .... *Ozophora*

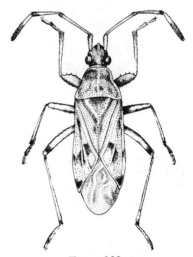

Figure 166

Figure 166 *Ozophora picturata* Uhler

This is a slender elongate long-legged species of a dull yellow color with dark brown patches and irregular spots present. The fourth antennal segment is white on the basal half. It is found on forest floors in shaded litter, particularly in oak-hickory woodlands. It is distributed across the southern and central states west to California and north to New England and Iowa. Eight additional species occur in the southern and southwestern states.

48a Anterior pronotal margin lacking a distinct collar 49

49 Lateral pronotal margins broadly explanate, the shelf-like explanation bearing numerous black punctures .... *Emblethis*

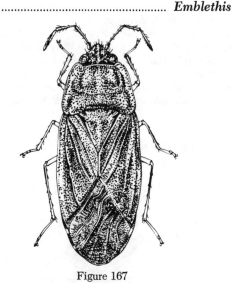

Figure 167

Figure 167 *Emblethis vicarius* Horvath

This is a rather large (6-8 mm) broadly ovoid pale yellowish-brown insect with the dorsal surface thickly covered with dark punctures. Specimens vary in color from dull grayish to light tan. It is often a common species in dry sandy areas both along the seashore and inland, often occurring in very hot dry areas with sparse litter present. The species ranges over almost the entire country but it is in need of careful study as a species complex may be represented.

49a Lateral pronotal margins narrowly explanate or carinate, but if explanate the flange never bearing numerous black spots 50

50 Very tiny species, not over 2.5 mm in length; lateral pronotal margins slightly carinate but never with an explanate flange .... *Antillocoris*

Antillocoris minutus (Bergroth)

This is the smallest North American lygaeid. It is of a dark reddish-brown color with the legs a light yellowish brown. It occurs on the ground and usually is found in forest litter. *A. minutus* is most frequently found beneath gray and white birch, but also occurs under hemlock and maple and in sphagnum bogs. It is a northern species occurring from eastern Canada south to New Jersey and Missouri.

A. pilosulus (Stål) is a very similar species distinguishable by having a dense coat of long yellowish hairs on the body. It largely replaces *minutus* in the southeastern states.

A third species occurs along the Gulf Coast from Florida to Texas.

50a Larger species, always more than 3 mm in length; lateral pronotal margins narrowly but distinctly explanate 51

51 Lateral pronotal margins sinuate, strongly concave in central area; both anterior and posterior pronotal lobes coarsely punctate; dorsal surface dull or at most subshining .... *Drymus*

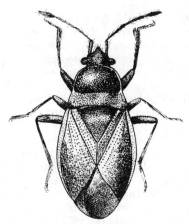

Figure 168

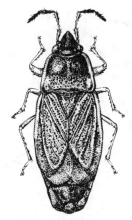

Figure 169

Figure 168 *Drymus unus* (Say)

This is a small elongate oval species (4-5 mm) of a dark brown color. The posterior pronotal lobe and the fore wings are often dull yellowish with dark brown markings. It is a common ground-living species in subclimax woodlands particularly where oak-hickory, black birch and red maple are intermixed, and it is most abundant along woodland margins. The species ranges widely throughout eastern North America south to North Carolina and west to Iowa and Texas.

D. crassus V. D. is a larger (5.5-7 mm) species of a very dark brown color throughout. It is found in more mature forests in leaf mold of beech-maple, hemlock, white birch and red spruce. It is known from New England south to North Carolina and west to South Dakota and Texas and is generally a scarce species.

51a Lateral pronotal margins evenly convex; surface strongly shining; anterior pronotal lobe impunctate .... *Cryphula*

Figure 169 *Cryphula trimaculata* (Distant)

This is a small (3-4 mm) dark reddish-brown species, often with a bluish iridescent cast to the body surface and with elongate yellowish stripes or spots on the scutellum. The first two antennal segments are dull yellowish, the third and fourth segments a contrasting dark brown to nearly black. The species is found in old fields with perennial bunch grasses and along woodland margins. It usually occurs on the ground in litter at the bases of grass clumps. *C. trimaculata* is a widespread not uncommon species from southern New England west to Colorado and south to Florida and Texas. Four additional species are present in the southwestern states.

REFERENCES
Slater, J. A. 1964. A catalogue of the Lygaeidae of the World. 2 vols. Univ. of Connecticut.
Sweet, M. H. 1964. The biology and ecology of the Rhyparochrominae of New England (Heteroptera: Lygaeidae). Pts. I & II. Entomol. Amer. 43:1-124; 44:1-201.

PYRRHOCORIDAE
Red Bugs or Cotton Stainers

This is a family of medium- to large-sized hemipterans usually marked with bright yellow, red, brown and white aposematic coloration. The lateral pronotal margins are somewhat expanded and flattened and often somewhat reflexed. The female sixth visible sternum is not cleft or split. The body lacks hair above. The head is inserted in the thorax to the eyes and the beak is elongate, extending onto the abdomen. Seven species, all in the genus *Dysdercus,* occur in the United States and all are confined to the extreme southern states.

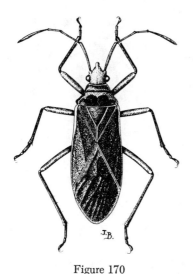

Figure 170

Figure 170 *Dysdercus suturellus* (H.-S.)

This is a large (11-17 mm) species. The head, base of the first antennal segment, femora, scutellum and the anterior pronotal lobe are bright red. The posterior lobe of the pronotum has a large dark brown quadrate patch. The wings are dark brown with pale yellow margins. The antennae, tibiae and tarsi are black. This and related species are known as Cotton Stainers or Red Bugs, and are often destructive to cotton. *D. suturellus* is also found on *Hibiscus* spp., oranges, *Solanum* and many other plants. It is common in Florida and adjacent southeastern states.

D. mimulus Hussey is a smaller species (7-10 mm) and somewhat similarly although usually less brightly colored. It has the first and second antennal segments nearly equal in length whereas in *suturellus* the first antennal segment is distinctly longer than the second. It occurs in the extreme southern United States from Florida to Arizona.

A third bright red species (*andreae*) with a prominent white "St. Andrews Cross" on the hemelytra occurs in southern Florida.

REFERENCE
Van Doesburg Jr., P. H. 1968. A Revision of the New World species of *Dysdercus* Guerin Meneville (Heteroptera, Pyrrhocoridae). Zool. Verhand. (Leiden) n. 97:1-215.

LARGIDAE
The Largid Bugs

This is a small family related to the seed bugs and cotton stainers. It is recognizable by the four-segmented beak, by the front wing membrane having seven to eight branching veins arising from two large basal cells, and by the absence of ocelli. Most species are stout-bodied and do not have the lateral pronotal margins expanded or bent upward and inward (reflexed). In the female the sixth visible abdominal sternum is split in the middle for its entire length. The various species are plant feeders found on the ground or on low weedy vegetation. Some are excellent ant mimics. Seven genera occur in the southern United States, three of which are confined to the southwestern states.

1 Eye produced on short thick stalks and placed near anterior margin of pronotum (fig. 171); pronotum not divided into two distinct lobes .... *Largus*

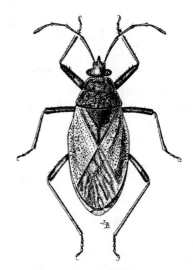

Figure 171

96 Families of Hemiptera

Figure 171 *Largus succinctus* (L.)

This large hemipteran (13-17 mm) is black or dark yellow-brown in color with the pronotal margins, lateral corial margins, edges of abdomen and basal portion of femora bright orange-red or orange-yellow. The body is thickly covered with short grayish hairs. The range is from southern New York south to Florida and west to Colorado and Arizona.

L. cinctus (H.-S.) is a closely related western species occurring throughout the southwest. There is need for further study to determine whether or not some of these "species" are actually distinct forms or merely geographic races. *L. cinctus* is said to differ in having the first segment of the beak less than twice the length of segment four (the reverse is true of *succinctus*). Six additional species occur in the southern and western states.

1a Eyes not produced on short thick stalks; head globular, eyes set well away from anterior pronotal margins; pronotum divided into two distinct lobes. Fig. 172 .... *Arhaphe*

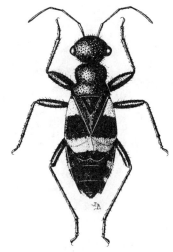

Figure 172

Figure 172 *Arhaphe carolina* H.-S.

This is a striking ant mimic (8-10 mm), jet black in color with short wings that reach only to the fourth abdominal segment and have the membrane reduced to a small flap. Each corium has a large triangular white patch along the outer half at the base, and an ovoid white patch at the apex. The head is very large and ovoid, giving the insect a distinctly ant-like appearance. A scarce species, it occurs in the southern states from North Carolina west to Arizona. Three additional species occur in the United States, one in Kansas and the others in the southwest.

REFERENCE

Halstead, T. F. 1972. A review of the genus Arhaphe Herrich-Schaffer (Hemiptera: Largidae). Pan. Pac. Ent. 48:(1):1-7.

Halstead, T. F. 1972. Notes and synonymy in Largus Hahn with a key to United States species (Hemiptera: Largidae) Pan-Pac. Ent. 48:246-248.

PIESMATIDAE
The Ash-Gray Leaf Bugs

This is a very small family whose members are recognizable by the two-segmented tarsi, the pronotum with five raised longitudinal ridges, the corium and clavus composed of an irregular network of cells (reticulated) and by the juga extending in front of the end of the tylus. Ocelli are present. Only one genus is found in North America.

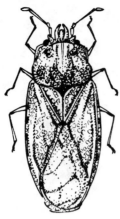

Figure 173

Figure 173 *Piesma cinerea* (Say)

P. cinerea is a small (2.5-3.5 mm) dull gray or yellow insect. The first two antennal segments are short, with the third as long as one and two together. The scutellum is very short and swollen at the tip. It is an abundant species usually collected by sweeping pigweed (*Amaranthus*), but reported from many other plants as well. The range is nearly throughout the United States.

Nine additional very closely related species have been described from the United States.

REFERENCE
McAtee, W. L. 1919. Key to the Nearctic species of Piesmatidae (Heteroptera). Bull. Brook. Ent. Soc. 14:80-93.

THAUMASTOCORIDAE
The Royal Palm Bugs

This is a very small family of which only a single species occurs in the United States. The body is very flattened dorso-ventrally, and the juga are parallel-sided and extend forward nearly to the end of the tylus, but in the American species do not exceed the tylus. The tarsi are two-segmented.

Figure 174 *Xylastodoris luteolus* Barber

This is a minute insect, 2-3 mm long, of a pale yellowish color with the apical half of the fourth antennal segment dark brown. The pronotum is strongly punctured with the lateral margins explanate and the wings semi-transparent. The males have an asymmetrical genital capsule that in some individuals curves to the right, and in others to the left. This is apparently an adaptation to facilitate mating between the closely appressed fronds of the host plant. These curious insects are seldom collected, as they live in the fronds of towering royal palms in subtropical southern Florida, where they sometimes cause serious injury to the plants.

REFERENCES
Baranowski, R. 1958. Notes on the biology of the royal palm bug, *Xylastodoris luteolus* Barber (Hemiptera, Thaumastocoridae). Ann. Ent. Soc. Am. 51:547-551.
Drake, C. J. & J. A. Slater. 1957. The phylogeny and systematics of the family Thaumastocoridae (Hemiptera: Heteroptera) Ann. Ent. Soc. Am. 50:353-370.

Figure 174

BERYTIDAE
The Stilt Bugs

The bugs of this small family are all very slender, elongate insects with long legs and antennae that give them a "thread-legged" appearance. Beginning students of Hemiptera sometimes confuse stilt bugs with marsh-treaders (Hydrometridae) or with thread-legged assassin bugs (Emesinae). The stilt bugs, however, never occur on the surface of water, and never have the front legs fitted for grasping prey. The labium and the antennae are both four-segmented. The first antennal segment is very elongate. Most of the common species are of a dull yellow-brown color but some of the less common species are very bizarre with various types of spines present over the body. The various species occur on vegetation or on the ground and are plant feeders. Seven genera occur in America north of Mexico. Many authors use the names of Neididae and Berytinidae for this family.

1   Head elongate with tylus produced into a bent down spine (horn) or a compressed plate (figs. 175, 176); underside of abdomen coarsely punctate; scent gland ostiole never produced into a spine-like process 2

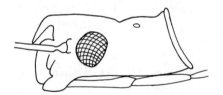

Figure 175

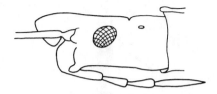

Figure 176

1a   Head short, tylus not produced forward as a spine or compressed plate, or if so then scent gland ostiole produced into a spine-like process (figs. 177, 178); underside of abdomen impunctate 3

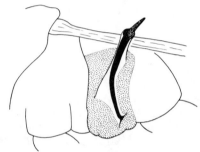

Figure 177

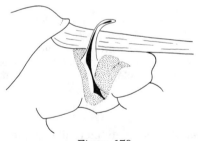

Figure 178

2   Head with a prominent down-curved cylindrical process produced between bases of antennae (fig. 175); hind femora extending backward to end of abdomen .... *Neides*

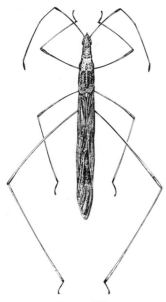

Figure 179

Figure 179 *Neides muticus* (Say)

This species (9-10 mm) is large for the family. It is pale dull yellow in color with the distal ends of the tibiae and tarsi and all of the fourth antennal segment dark brown to black. The labium extends back to the mesocoxae. The pronotum has the posterior 2/3 much more coarsely punctate than the anterior 1/3. *Muticus* is usually collected by sweeping in weedy fields. Nymphs have been taken on orange hawkweed. It is widely distributed over much of the United States but appears to be most common in the north. The southern limits of the distribution are poorly understood. This species has considerable superficial resemblance to species of the genus *Jalysus*.

2a Head with a forward produced and laterally flattened plate-like process (fig. 176), lacking a decurved spine-like process; hind femora short, not nearly extending to end of abdomen ...... *Berytinus*

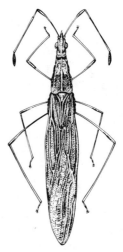

Figure 180

Figure 180 *Berytinus minor* H. S.

This is a small (6-7 mm) straw-colored species which has much shorter legs than most of the other stilt bugs. The labium extends back only to the fore coxae.

B. minor is a European species that apparently has only recently established itself in North America. It was first discovered in Michigan and has become rather common in southern New England and in New York where it is usually found on the ground around the base of grass clumps in old pastures and weedy fields. The potential spread of this introduced species will be of interest in years to come.

3 Pronotum bearing numerous elongate sharp spines on dorsal surface. Figs. 181, 182 4

3a Pronotum lacking numerous elongate sharp spines on dorsal surface 5

4 Clavus and corium bearing numerous elongate sharp spines along the veins (fig. 181); head and anterior pronotal lobe pale whitish yellow, concolorous with remainder of dorsal surface ....
.... *Acanthophysa*

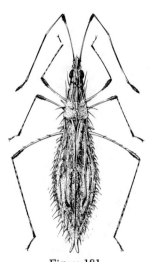

Figure 181

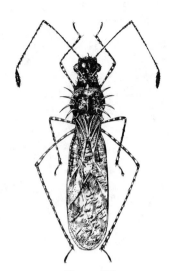

Figure 182

Figure 181 *Acanthophysa echinata* Uhler

This is a tiny stilt bug (3-4 mm) named *echinata* because of the numerous spines present on the body. In addition to the spines on the corium and pronotum it has a series of very long spines running in a median line along the tylus. It is pale yellowish white in color. The legs are ringed with dark bands. It occurs in both the short and long winged form. These are usually found on or around the bases of grass stems. The known distribution is confined to the Pacific coast states and to Arizona and Utah. A second species is known from the northwest.

4a   Clavus and corium lacking series of elongate sharp spines along the veins; head and anterior pronotal lobe shining black .... *Pronotacantha*

Figure 182 *Pronotacantha annulata* Uhler

This is a small (4 mm) bizarre species. The head and anterior pronotal lobe are polished black. (Frequently the posterior half of the posterior lobe is also shining black.) From the pronotum project a series of elongate sharp white spines that are often tipped with black. A circle of white globose projections occurs transversely around the base of the head. The hemelytra are transparent and the legs and antennae are conspicuously ringed with numerous dark red bands.

This striking little stilt bug is known in the southwestern states from Texas to California. It has been collected on yellow columbine and on *Antirrhinum* sp.

5   Fourth antennal segment longer than length of head; front coxae separated by a distinct sulcate area; scent gland channel groove lying on same side of lateral spine-like process for its entire length. Fig. 177 .... *Jalysus*

Families of Hemiptera

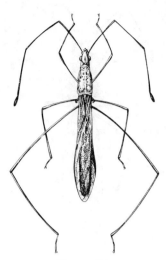

Figure 183

Figure 184

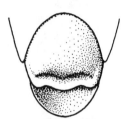

Figure 185

Figure 183 *Jalysus spinosus* (Say)

This is one of the commonest stilt bugs in the eastern United States, and its range extends westward to the Great Plains. It is of a bright brown color with the scent gland channel running along a sharp elongate spine (fig. 177). The pronotum is heavily punctured. The front wings are in large part semi-transparent. The species is most commonly found in fields and on shrubs. Food plants include *Oenothera biennis* L., species of *Gaura*, *Panicum* and tomatoes among others. It sometimes breeds on ornamental gourds (*Lagenaria*). It has recently been shown that *spinosus* is in part predatory, with survival much higher when animal food is available than without it.

J. wickhami V. D. is an extremely closely related species, so much so that they are frequently confused. The only certain method of distinguishing *wickhami* from *spinosus* is by the differently shaped genital capsule of the males which in *wickhami* has a thicker, heavier hind margin, a median ridge (faint) and lacks a transverse furrow (fig. 184) (present in *spinosus* [fig. 185]). The distribution extends from coast to coast over most of the country. *Wickhami* is frequently found in the same habitat as *spinosus* and they are often collected together. Three additional species occur in the southwestern states.

5a Fourth antennal segment shorter than length of head; front coxae contiguous or nearly contiguous; channel of scent gland strongly twisted to run along outer surface basally, then curving sharply to terminate along upper surface. Fig. 178 .... *Aknisus*

Aknisus multispinus Ashmead

This is a small (5-6 mm) straw-yellow or dull greenish-yellow stilt bug closely resembling members of the genus *Jalysus*. The distal ends of the tarsi and tibiae and all except the extreme tip of the fourth antennal segment are black. There is a small erect spine present between the antennal bases. It is usually collected around the bases of grass stems or swept from weedy fields. *A. multispinosus* is a scarce but widely distributed species that occurs throughout the southeastern United States north at least to New Jersey and west to Oregon and Arizona. The general range appears to be southern. One additional species occurs in California.

ARADIDAE
The Flat Bugs

This is a moderate-sized family of dorso-ventrally flattened, usually black, brown or chocolate-colored, insects. The body is generally elongately oval and the surface usually appears rather granular. Ocelli are never present.

These curious insects are usually found on or under the bark of dead trees where they feed upon fungi by means of extremely long thread-like stylets that are inserted into the tissues of the fungus. Most flat bugs require specialized collecting methods such as stripping bark and searching ground litter and are only occasionally taken by general sweeping, at lights and by other conventional methods.

1   Bucculae small, not extended strongly forward on either side of tylus to exceed it, nor forming a cleft apex; labium reaching or extending beyond fore coxae. Fig. 186 .... *Aradus*

This is a very large genus containing over 75 described North American species. This genus is a difficult place for the beginning student to attempt accurate species determination. The following key covers the most commonly taken species but represents only an introduction to a complex group of insects.

     a   Pronotal carinae on either side of midline absent or obsolete, at most only slightly developed anteriorly; antennae only very slightly longer than head .... *A. cinnamomeus* **Panzer**

This is a very small bright reddish-brown insect (3-5 mm) with very short stout antennae. The labium reaches onto the mesosternum. The reddish color is diagnostic. It is one of the few flat bugs that appears to feed on living trees. It has been reported several times damaging pines (*Pinus*). The species is widely distributed in North America and also occurs in Europe.

     aa   Carinae of pronotum on either side of midline distinct (fig. 186) and reaching or nearly reaching anterior margin; antennae distinctly longer than head ...... b

     b   Second and third antennal segments of nearly equal length, third segment never more than one-fifth shorter than segment two, both segments slender and evenly rounded .... *A. crenatus* **Say**

This is one of the largest species (8-11 mm) in the eastern United States. The abdominal connexivum is strongly "notched" (crenate) giving an irregularly scalloped appearance to the lateral margins. It is pale brown or brownish yellow, and has been taken under the bark of many kinds of trees. It is widely distributed east of the Mississippi River.

     bb   Second antennal segment more than one-fifth longer than third, one or both often swollen (fig. 186), not evenly cylindrical c

     c   Third antennal segment greatly enlarged, at least one-half thicker than segment two ...... *A. quadrilineatus* **Say**

This is a large (7-9 mm) dark reddish-brown species which is readily recognizable by the peculiarly enlarged third antennal segment, and by having the distal third of the second antennal segment yellow. The tibiae are dark brown with yellow distal and sub-basal bands. It occurs on oaks (*Quercus*). It is very widely distributed over most of the United States.

     cc   Third antennal segment not or only very slightly thicker than at least distal portion of segment two .... d

     d   Antennae very thick, at the thickest part considerably broader than fore femora. Fig. 186 .... *A. robustus* **Uhler**

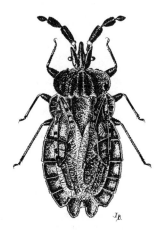

Figure 186

This is a small (5-7 mm) uniformly-colored brown to gray species with the head as long as the pronotum. It is easily recognizable by the large, thick second antennal segment which is covered with stubby semi-erect thick spines. It is usually found under the bark of red and black oaks. *A. robustus* is a very common species widely distributed over most of the country east of the Great Plains.

dd Antennae more slender, usually not thicker at their widest point than width of fore femora .... e

e Lateral margins of pronotum evenly rounded or granulate with very small teeth limited to areas near anterior and posterior angles; corium weakly dilated at base .... *A. lugubris* Fallen

A. lugubris is a rather small (5-6.5 mm) species, dark chocolate-brown in color with the distal third of the third antennal segment and the postero-lateral area of each abdominal connexivum whitish or yellowish. The species is found on *Pinus*, *Picea* and *Juniperus*. It is a Holarctic species which in North America ranges across the northern states and southern Canada from coast to coast.

ee Lateral margins of pronotum with small saw-like teeth present nearly throughout; corium always strongly dilated at base .... f

f Length second antennal segment about equal to or very slightly greater than interocular width; third antennal segment white or yellow with exception of base .... *A. similis* Say

This is a moderate-sized flat bug (5-8 mm) of a dark brownish black color. The short second antennal segment and particularly the pale third segment are characteristic. *A. similis* is associated particularly with shelf-fungi growing on birch trees. It is widely distributed in the eastern United States, south into Florida and west to Kansas and Texas.

ff Length second antennal segment subequal to width of head across eyes; third antennal segment nearly uniformly dark red-brown .... *A. acutus* Say

This species is recognizable by the uniformly dark chocolate-brown antennae, the long, distinctly clavate second antennal segment, and by the series of squarish granulate white spots laterally on the upper surface of the abdomen just within the broad connexivum. It occurs on oak (*Quercus*). The distribution encompasses most of the United States.

1a Bucculae large and elongate, usually extended forward beyond tip of tylus to either form a rather cleft head or to come into contact anterior to apex of tylus; labium not exceeding base of head 2

2 Scutellum rounded distally; labium arising from a cavity (atrium) which is open posteriorly. Fig. 187 .... *Aneurus*

Figure 187

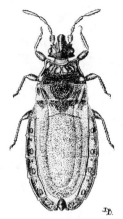

Figure 188

Figure 188 *Aneurus simplex* Uhler

This is a pale reddish-brown flat bug with translucent white wings. The length is about 5 to 5.5 mm. The spines on the antenniferous tubercles are acute. It is a northern species, probably ranging across the entire northern states and southern Canada.
  A. fiskei Heidemann is a smaller species (3.5-4 mm) differing from *simplex* in having the first antennal segment as long as the second (in *simplex* the first segment is distinctly shorter than the second). It is found in the eastern states west to Indiana. There are five additional North American species.

2a Scutellum triangular in shape; labium arising from a closed atrium. Fig. 189 3

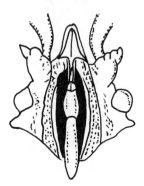

Figure 189

3 Abdominal sterna four, five and six with a narrow, sharp transverse ridge running just behind anterior margin of each sternite; third antennal segment at most only slightly longer than second .... *Neuroctenus*

Figure 190

Figure 190 *Neuroctenus simplex* (Uhler)

This is a blackish or very dark red-brown species 4.5-5 mm in length. The bucculae are rounded and meet in front of the tylus, but do not form a "cleft" at the front of the head. The membrane is whitish and semi-transparent, and the ends of the fourth antennal segments and the tarsi are yellow. It is usually found under the bark of various trees such as oak (*Quercus*) and beech (*Fagus*), but sometimes it is swept from weeds. *Simplex* is one of the commonest flat bugs throughout the eastern United States and occurs west to the Great Plains. Three additional species occur in the eastern United States.

3a No ridge present just behind anterior margins of abdominal sterna four, five and six; third antennal segment distinctly longer than second .... *Mezira*

Figure 191

Figure 191 *Mezira reducta* Van Duzee

This is a large species (7.5-8 mm) with the apical margin of the corium curving (sinuate) instead of being evenly rounded. The first antennal segment only slightly surpasses the end of the head, and is very strongly thickened on the outer two-thirds. This species is known only from California where it has the remarkable habit of living in termite galleries and presumably feeds on the fungus growing there.

M. lobata (Say) is similar in size to *reducta* but has the apical margin of the corium evenly rounded instead of sinuate, and it has the first antennal segment not strongly thickened and greatly exceeding the end of the head. It is usually taken under bark. It is widely distributed nearly throughout the United States. There are five additional North American species.

REFERENCES
Parshley, H. M. 1921. Essay on the American species of Aradus (Hemiptera). Trans. Amer. Ent. Soc. 47:1-106.
Usinger, R. L. & R. Matsuda. 1959. Classification of the Aradidae (Hemiptera-Heteroptera). British Museum (Nat. Hist.) London.

TINGIDAE
The Lace Bugs

This is a large family of small, often attractively shaped and ornamented species. The common name refers to the numerous small lace-like cells in the expanded pronotum and fore wings.

The family is characterized by the lack of ocelli; wings with numerous cells and without distinct divisions into clavus, corium and membrane; antennae four-segmented with the first two segments short and stout and the third most frequently very elongate and slender (but sometimes thickened and enlarged); tarsi two-segmented.

All of the lace bugs are plant feeders, and often a species is restricted in its feeding habits to a single host plant or a single group of related plants. Although most genera occur on herbaceous growth, some of the commonest North American species feed on the leaves of trees. Some species are destructive.

1 Pronotum with two swollen bulbs on midline, posterior one much larger than anterior .... *Galeatus*

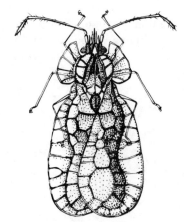

Figure 192

Figure 192 *Galeatus spinifrons* (Fallen)

This is a beautiful lace bug (4-4.5 mm) with a very transparent appearance. The head has 5 large elongate sharp black spines and the paranota are very large and broad with a single row of 5 cells. The front wings are also very broad with the costal area composed of a single series of large rectangular cells. It has been taken on *Aster macrophyllum, Eupatorium, Artemisia, Chrysanthemum* and a number of related plants. The range is primarily northern from the northeastern states west to Alberta, Colorado, Utah and New Mexico.

1a Pronotum with one or no bulbs on midline, posterior bulb always absent 2

2 Paranota present, with sharp spines present on lateral margins. Fig. 193 .... *Corythucha*

This is the largest genus of lace bugs in North America. Nearly 50 species occur north of Mexico. The various species are typically "lacy" forms with broad many-celled paranota and front wings. The body form is usually rectangular with a prominent globular "hood" on the anterior part of the pronotum that extends forward to usually completely cover the head. Many species are common and widespread and the majority live on the leaves of trees and shrubs. Identification is often greatly facilitated if the collector is careful to note on what plant his specimens were collected. The following key will aid in identification of the more common and widespread species but sometimes other species will be found locally abundant on particular food plants. As with several other large genera treated in this book the keys to species are only introductions and the serious student should turn to primary literature sources for definitive identification.

a Pronotal hood very large, appearing to be nearly three times as high as median carina .... b

aa Pronotal hood never more than twice as high as median carina .... e

b Entire lateral margins of both paranota and fore wing completely dark. Fig. 193 .... *C. bulbosa* O. & D.

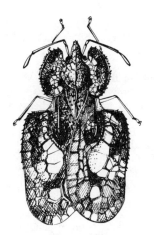

Figure 193

This is the largest species of the genus being usually about 4.5 mm. It is readily recognizable by the completely dark lateral margins of the paranota and hemelytra which gives the species a nearly black appearance. The large hood is very sharply bent downward so that its front margin is almost vertical to the longitudinal axis of the body when viewed from the side. The host plant is the bladder nut (*Staphylea trifolia*). It is generally not a common insect and is recorded through the central part of the eastern and midwestern states from Pennsylvania, New Jersey and Virginia, west to Illinois and Missouri.

bb Costal area of fore wings usually with a dark basal and pre-apical crossbar, middle portion of lateral margin always pale c

c Posterior margin of pre-apical dark crossbar on fore wings straight .... *C. cydoniae* (Fitch)

This is a small species (3.5 mm), often abundant on hawthorn (*Crateagus*), quince (*Cotoneaster*), chokeberry (*Pyrus arbutifolia*) and allied plants. The pronotal hood is sharply compressed anteriorly and the costal wing margins have very short spines. This is a very widely distributed species occurring nearly throughout the United States.

cc Posterior margin of pre-apical dark crossbar on front wings rounded or irregular .... d d   Length over 4 mm; hood of pronotum very large and as wide as high .... *C. associata* **O. & D.**

The paranota of this rather large species have one or two small brown spots. The median carina of the pronotum is arched and bears two rows of cells. Its hosts are wild cherry (*Prunus serotina* and *P. persica*). The range is nearly throughout the eastern United States west to Iowa.

C. distincta O. & D. will also run here in the key. It is a common species on *Carduus lanceolatus, Cnicus* sp., *Lathyrus nuttalli* & Hollyhock in the northwestern United States.

dd Less than 4 mm in length; hood of pronotum distinctly higher than wide .... *C. pallipes* **Parshley**

The median carina of the pronotum is highest anteriorly with the anterior cell very large and the posterior part of the hood very large and globose. This lace bug lives chiefly on birch (*Betula*), maple (*Acer*), and willow (*Salix*) but is also found on other trees. It is distributed across the northern states and adjacent areas of southern Canada.

e   Dorsal surface almost uniformly white, lacking both basal and pre-apical dark crossbars on front wings but having a dark mark at apex of discoidal area. Fig. 194 .... *C. ciliata* (Say)

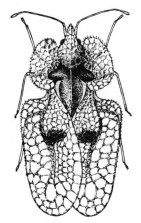

Figure 194

This is an easily recognized species because of its nearly uniformly milky coloration. The dark discoidal mark is diagnostic. The hood is very low, only slightly higher than the median carina. It is an abundant species on sycamore (*Platanus occidentalis*), often occurring in such numbers as to cause the leaves to whiten and drop prematurely. It has also been reported on ash (*Fraxinus*), hickory (*Carya*), mulberry (*Morus*) and several other plants. The species occurs over much of the eastern and central United States west to the Rocky Mountains.

ee Dorsal surface of fore wings with dark basal crossbar present, or with a series of numerous brown spots, usually with a pre-apical dark crossbar as well (if largely milky-white, then most cells partially opaque rather than transparent) .... f

f   Lateral margins of paranota almost devoid of spines ...... *C. mollicula* **O. & D.**

In this species the median pronotal carina is low and only very slightly arched. The posterior crossbar is often interrupted on the midline. This species feeds on willow (*Salix*), sometimes on poplar (*Populus*) and is found over much of the United States and southern Canada east of the Rocky Mountains and in the Pacific Northwest.

ff Median lateral margins of paranota bearing a row of distinct spines .... g

g   Front wings usually with numerous brown (often faint) spots but these not forming distinct dark basal and pre-apical crossbars; sometimes nearly uniformly white but in this case most cells of wings usually opaque milky-white with only center of each cell transparent. Fig. 195 ...... *C. marmorata* (Uhler)

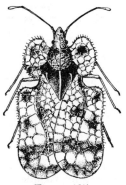

Figure 195

This is a small (3.2-3.4 mm) very pale species easily recognizable by its largely opaque whitish ground color. The brown spots are usually vague and ill-defined. It is one of the most abundant of the lace bugs, and occurs on various Compositae in fields, chiefly goldenrods (*Solidago*) and asters (*Aster*), and sometimes causes severe injury to chrysanthemums. The range is almost throughout the United States and southern Canada.

C. gossypii (F.) also keys here. It usually has the cells less distinctly suffused, and may readily be recognized by having the median pronotal carina as high as the hood. It feeds on a great variety of plants including many cultivated crops. The distribution is southern from Florida at least to Texas.

gg Front wings always with a distinct dark basal crossbar; cells of wings usually either entirely transparent or darkened .... h

h Front wings with only basal crossbar, pre-apical bar at most represented by a trace .... *C. ulmi* **O. & D.**

This is a small species (3.3 mm) found on elm (*Ulmus*) (some specimens of *arcuata* will key here). It is only occasionally common despite its wide distribution from New England south to South Carolina and Mississippi and west to South Dakota and Nebraska.

hh Front wings with both basal and pre-apical dark crossbars i

i Hood low, only slightly higher than median carina .... j

ii Hood considerably higher, nearly twice as high a median carina .... k

j Length over 4 mm; "pre-apical" dark crossbar of front wings well-developed and nearly black; pronotal hood, in dorsal view, only moderately constricted on tapering anterior portion; occurs on wild cherry .... *C. pruni* **O. & D.**

The coloration of this large common species is strongly contrasting. The lateral margins of the paranota and wings have only a few short spines. The length is from 4.0-4.2 mm. Its host is wild cherry (*Prunus serotina*). It is widely distributed throughout the eastern and central United States with western records from Utah and Oregon.

jj Length less than 3.5 mm; pre-apical dark crossbar on front wings much paler; anterior portion of hood sharply and abruptly constricted; occurs on oaks .... *C. arcuata* (Say)

This species looks much like a small pale *C. pruni*. In some specimens the pre-apical crossbar is obsolete or absent. The median carina is arched anteriorly and has two rows of cells. It occurs in abundance on various species of oaks (*Quercus*) and occasionally on rose (*Rosa*), maple (*Acer*), apple (*Malus*) and chestnut (*Castanea*), and is found throughout the United States where these species occur.

k Hood twice as long as high .... *C. pergandei* **Heid.**

This is a very small species (2.8-3.0 mm) with the darkened bars of the front wings anteriorly pale brown. The marginal spines are numerous but very short. The species occurs mostly on alder (*Alnus*) but is recorded also on hazel (*Corylus*), elm (*Ulmus*), birch (*Betula*), crab apple (*Malus*) and other species of plants throughout most of the United States.

kk Hood less than twice as long as high .... *C. juglandis* (**Fitch**)

This species occurs on walnut (*Juglans nigra*), butternut (*J. cinerea*), linden (*Tilia* sp.) and other deciduous trees throughout the eastern and central states west to Kansas and Texas.

C. padi Drake is a very similar species that is common on *Prunus demissa* in the northwestern United States.

2a Paranota either present or absent; when present, without marginal spines (fine elongate ciliate hairs may be present) .. 3

Figure 196

3   Paranota excavated or with a basal fold opposite calli (fig. 196) along anterior half of inner margin; hood entirely covering head and extending anteriorly to or beyond second antennal segment .... *Corythaica*

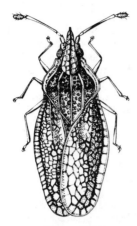

Figure 197

Figure 197   *Corythaica bellula* Torre-Bueno

This is a tiny lace bug (1.9-2.2 mm) that somewhat resembles small species of *Corythucha* by having a long narrow hood, but *C. bellula* lacks marginal paranotal spines. The upper surface is a dull greyish white. The median line of the hood and the area of the front wing outside the discoidal area but not including the marginal row of cells are dark. It occurs in long winged (rare) and relatively short-winged forms. The food plant is *Panicum* grass. At present this species is known from New England, New York, Florida and Nebraska.

C. venusta (Champ.) is a much larger, ovoid species. It is easily distinguished from *bellula* by the larger hood, strongly flaring paranota and the presence of 3 conspicuous dark patches along the lateral margins of the front wings. The host plants are *Eriogonum* sp., *Lantana* sp. and *Salsola pestifer*. It is common and widely distributed in the western half of the United States and north into Canada. Two additional species occur in the western states, one of which also occurs in Florida.

3a   Paranota if present lacking anterior excavated area, or if excavated area present hood not extending anteriorly beyond head 4

4   Paranota folded inward over pronotum (reflexed) so that true lateral margins are innermost and in contact with pronotal carinae. Fig. 199 5

4a   Paranota present or absent, when present if folded inward and reflexed against pronotum then not reaching carinae 6

Figure 198

5   Paranota longitudinally creased, outer half folded flat against inner half. Fig. 198 .... *Leptodictya*

Leptodictya plana Heid.

This is a rather elongate, elliptical lace bug with a very small hood and three pronotal carinae. The peculiar expanded but reflexed paranota are distinctive. The species is found on grasses and restricted in distribution to the southern states from Florida west to Kansas and Texas.

L. simulans Heid. has the sides of the paranota weakly rounded instead of being straight as they are in *plana*. It occurs on bamboo. *Simulans* has a distribution similar to that of *plana* but ranges in the east as far north as Virginia. Three additional species occur in the Southwest.

5a   Paranota folded back over pronotum from base, not forming a sharp longitudinal crease .... *Physatocheila*

Physatocheila plexa (Say)

This is a uniformly reddish or yellowish-brown species, 3-3.5 mm long. The hoodlike pronotal elevation and the reflexed paranota are coarsely punctured. The body is elliptical in shape and the head bears five curved spines. *P. plexa* occurs on oak (*Quercus*), hickory (*Carya*), willow (*Salix*), and mountain laurel (*Kalmia latifolia*). It is distributed widely in the eastern states and eastern Canada from Nova Scotia and New England south to North Carolina and Tennessee and west to the Great Plains.

P. major O. & D. is a larger species, 4.5 mm or more in length, with the third antennal segment enlarged from the proximal to the distal end instead of being linear as it is in *plexa* and *brevirostris*. The known range is from Maryland and Virginia west to Illinois.

P. brevirostris O. & D. is similar in size and color to *plexa* but the labium extends backward only between the middle coxae, whereas in *plexa* it extends to between the hind coxae. The range is from Quebec and New England south to Virginia and west to Illinois and Iowa.

Figure 200

Figure 199

P. variegata Parshley (fig. 199) is 3.4-4 mm in length, of a greyish-brown color marked with dark brown that gives it a light and dark, mottled appearance. The labium is very long, usually reaching the second visible abdominal segment. Little is known of its food plants but it has been taken on hickory (*Carya*), willow (*Salix*), alder (*Alnus*), cottonwood (*Populus*) and pine (*Pinus*). The species ranges from New England west to Iowa and Missouri and south to Virginia. Variety *ornata* V.D. is found in California, Oregon and Idaho.

6 Channel for reception of labium interrupted at suture between the mesosternum and metasternum by a transverse plate. Fig. 200 .... *Gargaphia*

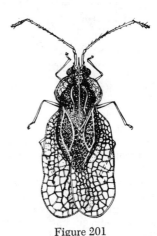

Figure 201

Figure 201 *Gargaphia tiliae* (Walsh)

This is a rather large lacy tingid (4-5 mm) largely of a pale white color. The cells of the front wings are large and transparent; the head has five spines present but when viewed from above appears to have only three. The first three antennal segments are yellow. The lacy appearance gives it a resemblance to species of *Corythucha*. *G. tiliae*, often abundant on linden (*Tilia*), occurs throughout the eastern United States and west to Nebraska, Kansas, Colorado and Arizona.

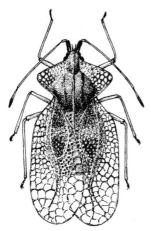

Figure 202

G. *solani* Heid. (fig. 202) is known as the "eggplant lace bug" as it often becomes destructive to that plant, but also occurs on other members of the family Solanaceae as well as other plants. It can be distinguished by its black or dark brown first antennal segment and the strongly angulate or narrowly rounded paranota. The pronotum is covered with long silky hairs. This and at least one other species of *Gargaphia* have the unusual habit of "guarding" the eggs and young nymphs, and actively guiding the movements of the latter. The distribution is chiefly southern from Maryland west to Arizona, but it is also recorded from Pennsylvania, New Jersey, Connecticut and British Columbia.

G. *angulata* Heidemann is a small species, not over 3.5 mm, that closely resembles G. *solani* in appearance. The head spines are very short and not visible in front of the hood. The lateral margins of the paranota are distinctly angulate as in *solani* rather than rounded. The first antennal segment is pale. It occurs chiefly on Jersey Tea (*Ceanothus americanus*), but sometimes injures beans. The range is from New England south to the Gulf and west to Colorado and Arizona. Twelve additional species occur in the United States, chiefly in the southern and southwestern states.

6a	Channel for reception of labium not interrupted by a transverse plate	7
7	Paranota either lacking, costate, or narrow and reflexed subvertically	8
7a	Paranota explanate or somewhat broadly reflexed but never reflexed inward and flattened on dorsal surface of pronotum	13
8	Lateral carinae of pronotum absent on anterior two-thirds of pronotal disc and sometimes on posterior process	9
8a	Three well-developed carinae present on pronotum	10
9	Lateral carinae absent even on posterior process of pronotum, hemelytra narrow and elongate with lateral margins sinuate; antennae short and thick .... *Leptoypha*	

Figure 203

Figure 203 *Leptoypha mutica* (Say)

This is a small elongate dark reddish-brown to blackish, nearly parallel-sided tingid, length 2.7-2.9 mm. Only the median carina is present on the pronotum; a whitish "bloom" is often present on the head and the anterior part of the pronotum. The antennae are short and thick. It occurs on the foliage of ash (*Fraxinus*), the fringe tree (*Chionanthus virginicus*) and swamp privet (*Adelia acuminata*). The range is from Quebec and New England west to North Dakota and south to Florida and Texas. The species as currently understood may well be composite.

L. costata Parshley is similar in size but considerably broader in outline and readily distinguishable by having two rows of cells in the costal area of the wing near the base and a well-defined single series of cells from there all the way to the apex. In *mutica* cells are indistinguishable in this area on the basal half of the margin. The principal host plant is ash (*Fraxinus*). The range does not reach as far north as does that of *mutica*, extending from Connecticut south to the Gulf and west to Arkansas and Colorado.

There are seven or eight additional species in the United States, known mostly from the southern and southwestern states.

9a Lateral carinae present on posterior third of pronotal disc; hemelytra ovoid, their lateral margins evenly, arcuately convex; antennae slender .... **Dictyla**

Dictyla echii (Schrk.)

This elliptical species is easily recognized by the black head and pronotum with strongly contrasting yellow paranota that are broadly reflexed back over the pronotum. The femora, except for the distal ends and the tarsi, are very dark brown and contrast sharply with the yellow tibiae. In Europe it lives on many species of *Echium* and other related plants. This is a Palearctic species now known to occur in Pennsylvania, Maryland, West Virginia and Florida.

10 Anterior margin of pronotum convex; front wings elongate and narrow, becoming somewhat constricted posterior to end of discoidal area; pronotal carinae slightly closer to each other on posterior process than on disc .... **Teleonemia**

Figure 204

Figure 204 *Teleonemia nigrina* Champion

This is an elongate, relatively slender and nearly parallel-sided species, length 3.2-3.6 mm. It is of a dark grey-brown color and the fore wings possess a few black spots. The antennae are short and heavy. The female has a projecting spine from either side of the last abdominal segment. *T. nigrina* somewhat resembles *Leptoypha mutica* in shape. It occurs chiefly on *Eriogonum* and has been reported on a number of other plants. This is a southern species known to occur north to Virginia, New Jersey and Iowa and westward to Utah and California.

Thirteen additional species have been reported from the southern and western states.

10a Anterior margin of pronotum concave or straight; front wings either oval, elliptical or narrow; pronotal carinae parallel to one another 11

11 Third antennal segment extremely large and of almost even thickness throughout, if slightly tapered then labium not exceeding mesocoxae and discoidal area of fore wing poorly differentiated .... *Alveotingis*

Figure 205 *Alveotingis grossocerata* O. & D.

This is a handsome bizarre species (2.8-3.5 mm in length). The shape is ovoid with strongly convex rather beetlelike wings in the coleopteroid form. It is readily recognized by its enormously enlarged antennae with the thick, heavy third segment covered with coarse black hairs. The color is dark brown to nearly blackish with pale white hemelytral cells. Both macropterous and coleopteroid forms occur. The appearance is very similar to that of *Hesperotingis antennata*. It is a rare species known only from New England, New York, Pennsylvania, Iowa, and Kansas. Two additional species occur in the central United States.

11a   Third antennal segment sometimes very large but always broader distally than proximally 12

12   Third antennal segment very large, gradually increasing in width so that entire outer half is enlarged; antennae relatively short and stout, much shorter than body length .... *Hesperotingis*

Hesperotingis antennata Parshley

This is an elliptical medium-sized brown species, reaching a length of 4.5 mm. The large clavate third antennal segment is distinctive and resembles that of *Alveotingis* but is distinctly narrowed on the basal one-third. The labium reaches the hind coxae. The wing development is very variable, specimens occurring with long, coleopteroid and intermediate wing conditions. Little is known of the habits of this scarce species but it has usually been collected on or near the ground. The distribution is poorly known but it has been taken in various eastern states from New England to Florida and in Missouri.

H. occidentalis Drake is a bright yellowish-tan species. It may easily be distinguished from *antennata* by having the pronotum and hemelytra uniformly pale (darkened in *antennata*, particularly on the pronotum), and by having a much longer third antennal segment which is more than twice as long as the pronotum measured from the humeral angles to the anterior margin. In *antennata* the two are subequal. The host plant appears to be unknown. It is a fairly common species in the northwestern states from Colorado to California.

Five additional species are known, the majority with a southern or western distribution.

12a   Third antennal segment slender for most of its length, slightly enlarged at darkened distal end; antennae relatively elongate and slender, as long as or nearly as long as body length .... *Melanorhopala*

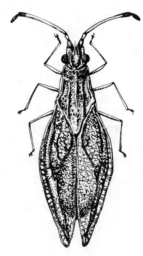

Figure 206

Figure 206 *Melanorhopala clavata* (Stål)

This is a large, elongate, nearly parallel-sided (macropterous form) brown lace bug from 4 to nearly 7 mm in length. The enlarged distal end of the third antennal segment and the entire fourth segment are dark chocolate-brown. The cells of the wing are numerous and small, forming a reticulate pattern. A number of different conditions of wing development occur and there is also marked sexual dimorphism. This is a common species usually taken in weed grown fields and often associated with goldenrod (*Solidago*). It is chiefly northern in distribution ranging from New England west to Manitoba and Wyoming, but it also has been reported from Missouri and Florida.

Two additional species occur in the United States, one from Virginia and Maryland, the other from Colorado.

13   Head without spines but sometimes with grooves and ridges present (the antenniferous tubercles may sometimes be rather spinelike) .... *Atheas*

Atheas insignis Heidemann

This is a small blackish rather parallel-sided species (2.3-2.5 mm long) with the first, second, and fourth antennal segments black, strongly contrasting with the yellow third antennal segment which is black only at the extreme base. The black coloration of the body is often obscured by a bluish-gray bloom. It lives on the pencil flower (*Stylosanthes biflora*). The range is in the southeastern states north to Maryland and west to Mississippi.

Figure 207

A. mimeticus Heid. (fig. 207) is similar in size and color but is distinguishable by having the first, second and entire proximal third of the third antennal segment black. It occurs on prairie clover (*Petalostemum purpureum*). The range extends further north than does that of *insignis* and the species is probably found throughout the southeastern states north in the Midwest to Wisconsin and Minnesota and west to Wyoming and New Mexico. Three additional species occur in the southern United States but the genus needs careful study as considerable variability is evident.

13a Head with definite spines present (sometimes difficult to see in specimens with large hoods) 14

14   Head mesally completely covered dorsally by membranous pronotal hood .... *Stephanitis*

Stephanitis rhododendri Horvath

This is a medium-sized species (3-3.7 mm) with a large hood. It somewhat resembles species of *Corythucha*. The paranota are wide, large and strongly bent upward. The antennae are very long and slender. The wings are very large, rounded at the end, and widen in a broad arc from the base to the tip. This is a native species that lives on rhododendron (*Rhododendron* sp.) and mountain laurel (*Kalmia latifolia*), and is often a pest on cultivated plants growing in open sunny places. It is distributed from New England to Florida and west at least to Ohio and also known from Oregon and Washington.

S. pyrioides (Scott) is a slightly larger species (up to 4 mm in length) and has a much larger and more globose hood than does *rhododendri*. It is pale in color and the hood and median carina are of about the same height. This species was introduced from Japan in shipments of azalea and now occurs in some abundance on the eastern seaboard from New England to Virginia and has also been reported from Missouri and Florida. It is frequently destructive to cultivated azaleas and rhododendrons.

S. takeyai Drake and Maa is another Japanese species that was first reported from North America in 1950 (but records go back to 1945).

It is now known from Connecticut and New Jersey. The insect is destructive to *Pieris japonica* in Connecticut but, as this plant is widely used in foundation plantings, can be expected to spread over a much greater geographic area. The entire hood is black and very large and globose, nearly twice the height of the median carina. One additional species occurs in Florida.

14a Hood when present relatively small, not completely covering head 15

15 Metathoracic scent gland openings present and distinct; body elongate, slender and nearly parallel-sided or sinuate .... *Leptopharsa*

Leptopharsa clitoriae (Heid.)

This is an elongate, subparallel-sided species of small size (2-2.3 mm long) with the lateral margins of the fore wings rather broadly rounded before the middle in dorsal view and crossed by a prominent black bar. The third antennal segment is shorter than the pronotal length. The hood is small, compressed and covers only the base of the head. The head spines are blunt. It occurs on butterfly pea (*Clitoria mariana*), tick trefoils (*Meibomia*), stickseed (*Lappula*) and bush clover (*Lespedeza*). The distribution is from New England south to South Carolina and west to Texas and Nebraska.

Figure 208

L. heidemanni (O. & D.) (fig. 208) is slightly larger (3-3.2 mm), lacks the dark subbasal crossbar, and has the third antennal segment as long as the pronotum. The greater part of the pronotum is black and there is a large dark spot just before the end of the discoidal area of the front wing. The head spines are yellow with dark tips and are sharp and needle-like. The food plant is wild indigo (*Baptisia tinctoria*). It ranges from New England west to Missouri and Arkansas and south to Louisiana.

L. oblonga (Say) closely resembles *heidemanni* but is usually a little smaller (2.7-2.8 mm), has white- instead of dark-tipped head spines, and has four instead of three cells along the median ridge of the hood. The food plant is *Amphicarpa bracteata*. The range is from New England west to the Dakotas and south to Virginia and Arkansas. Four additional species are known from this country.

15a Metathoracic scent gland openings absent; body broad and subelliptical .... *Acalypta*

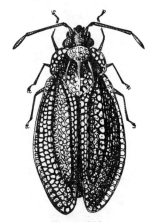

Figure 209

Figure 209 *Acalypta lillianis* Torre Bueno

This is a small ovoid grayish-black tingid with short antennae and stout but not appressed head spines. The dorsal surface is glossy. The pronotum has three carinae, the hood is small and extends only onto the base of the head. The members of this genus live among mosses on the ground. *A. lillianis* ranges widely from New England and Newfoundland west to British Columbia and Alaska and south to North Carolina and Iowa.

Eight additional species have been described from Canada and the United States; the majority are western although many are rare and known only from widely separated localities.

REFERENCES

Bailey, N. S. 1951. The Tingoidea of New England and their biology. Ent. Amer. 31:1-140.

Drake, C. J. and F. A. Ruhoff. 1965. Lace bugs of the world: A catalog (Hemiptera: Tingidae) U.S.N.M. Bull. 243:(i)-viii + 634 pp.

Hurd, M. P. 1946. Generic classification of North American Tingoidea (Hemiptera: Heteroptera) Iowa State Coll. Journ. Sci. 20:(4):429-492.

ENICOCEPHALIDAE
The Unique-Headed Bugs

These small insects (2-5 mm) are uncommon in North America. The family is readily recognizable by the unusual head which is constricted at its base and just behind the eyes, and somewhat swollen between these constrictions. The front legs are raptorial, the tarsi being capable of closing upon the end of the rather broad tibiae. Enicocephalids are generally found under rocks and leaves in wooded areas and under the bark of dead trees where they feed upon other insects. Some species form large swarms and fly about in the sunlight, resembling swarms of midges.

Three genera are known from the United States.

Figure 210

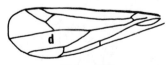

Figure 211

1 Discal cell of front wing closed distally (fig. 210); ocelli located on distinct elevations; posterior margin of pronotum truncate, not emarginate .... *Hymenocoris*

Hymenocoris formicina Uhler

This is a long, slender insect, clothed with more or less scattered long pale hairs; length 5.0 mm. It is known only from California.

1a Discal cell open distally (fig. 211); ocelli not located on distinct elevations; posterior margin of pronotum shallowly but distinctly emarginate .... *Systelloderes*

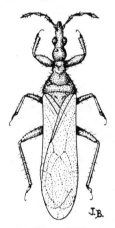

Figure 212

Figure 212 *Systelloderes biceps* (Say)

This is a rather small (3-4 mm) fragile insect with a more or less smooth body surface. It is pale yellow or grayish brown in color. The eyes are large and coarsely granulated. It ranges from New England south to Florida and west to Utah and Arizona.

One additional species is known from California and four others from Utah, Mississippi and Iowa.

REFERENCE

Usinger, R. L. 1945. Classification of the Enicocephalidae (Hemiptera, Reduvioidea). Ann. Ent. Soc. Amer. 38: (3):321-342.

PHYMATIDAE
The Ambush Bugs

This is a small family with relatively few species in North America. Some species are common insects usually found hidden in flower heads. They are very cryptically colored and wait motionless for small bees, flies and wasps which are captured by a rapid movement of the beautifully modified raptorial front legs and then the body contents of the prey are sucked out. Ambush bugs are stout-bodied, medium-sized insects that vary in color from yellow and brown, or black, to pale green with darker markings. The fore femora are greatly enlarged. The tibiae are sickle-shaped and fit into a groove on the inner surface of the femora; the tarsi are either very reduced or absent. Some authors (probably correctly) consider the ambush bugs as a subfamily of the Reduviidae.

1 Scutellum triangular, short, not as long as length of pronotum (fig. 213); anterior tarsi small but distinct .... *Phymata*

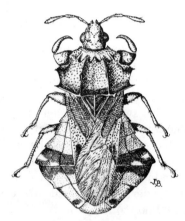

Figure 213

Figure 213 *Phymata pennsylvanica* Handlirsch

A moderate-sized (8.5-9.5 mm) ambush bug, usually of a yellow and brown or yellow and black color, although living specimens are often bright green and black. The membrane is brown. In males the fourth antennal segment is longer than the second and third antennal segments together, and the abdominal connexivum is abruptly and acutely expanded at the fourth connexival segment (figs. 213, 214).

Figure 214

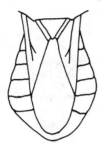

Figure 215

The distribution is throughout the eastern United States westward at least to Illinois.

P. americana Melin is a very similarly colored and shaped species. The males differ from *pennsylvanica* in having the fourth antennal segment subequal to or shorter than segments two and three combined and the abdominal connexivum is rather evenly expanded to the fourth connexival segment (fig. 215). This species, distributed throughout the United States, is segregated into several subspecies.

Approximately seventeen additional species occur north of Mexico, the majority being restricted in distribution to the southern and western states. Many of the species are very similar and accurate determination is critical.

1a   Scutellum not triangular, narrowly U-shaped, reaching to posterior end of abdomen (fig. 216); anterior tarsi absent .... *Macrocephalus*

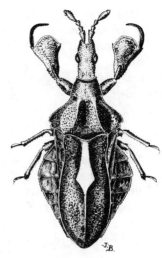

Figure 216

Figure 216   *Macrocephalus cimicoides* Swederus

A relatively large (9-11 mm) species. It is dull reddish yellow in color with the posterior pronotal lobe and the scutellum brown or reddish purple. The carinae on the posterior pronotal lobe are well developed and the humeral pronotal angles notched. The range is across the southern United States from the Carolinas and Florida to Texas and California.

M. prehensilis (F.) is a considerably smaller (5-5.5 mm) species. The pronotal carinae are obsolete and the pronotal humeral angles are entire and obtuse. It ranges from North Carolina south to Florida and west to Kansas and Texas.

At least six additional species are known from the southwestern United States.

REFERENCE

Kormilev, N. 1960. Revision of Phymatinae (Hemiptera, Phymatidae). Philippine Jour. Sci. 89:(3-4):287-486.

REDUVIIDAE
The Assassin And Thread-Legged Bugs

This is a large family of predatory hemipterans. Most species are active predators on other insects and several are of considerable economic value in reducing the numbers of destructive insect species. However, a few have developed the habit of feeding on vertebrate blood. These species are known as "kissing bugs" or "cone-noses" and several have been implicated in the transmission of a serious trypanosome disease to man. Many species "bite" readily, causing a very painful reaction, and they should be handled with caution.

The species vary greatly in habitus from stout robust insects to extremely elongate slender stick-like or "thread-like" forms. Despite their predatory habits the fore femora are generally slender and not obviously raptorial in appearance. The definitive recognition character for the family is the presence of a median longitudinal groove on the prosternum that is minutely transversely striated. The labium is stout, three-segmented, with the first segment generally curved outward away from the head.

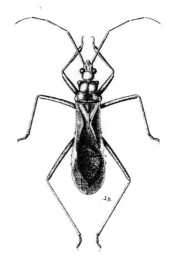

Figure 217

1   Anterior coxae never more than three times as long as wide; body shape variable, usually not extremely slender (assassin bugs) 2

1a Anterior coxae elongate, at least four times as long as wide; body extremely slender and "stick-like" (thread-legged bugs). Fig. 246 26

2   Ocelli absent 3

2a Ocelli present 4

3   Pronotum unarmed; apex of scutellum produced into a long horizontal tapering spine. Fig. 217 .... *Oncerotrachelus*

Figure 217 *Oncerotrachelus acuminatus* (Say)

This is a small elongate-oval species (4-7 mm) which is widest behind the middle of the abdomen. It is of a yellowish to dull reddish-brown color. The hemelytra mesally, the anterior third of the scutellum and the mesal area of the hind lobe of the pronotum are dark brown. The body and legs are thickly clothed with rather elongate fine erect hairs. The anterior pronotal lobe is swollen with a conspicuous median longitudinal groove present. It occurs mainly in debris in moist places. The distribution is over much of the United States east of the Rocky Mountains.

One additional species is found in Texas.

3a Posterior lobe of pronotum bearing a pair of elongate upward projecting spines midway along lateral margins (fig. 218); scutellum with an apical spine and a conspicuous upward projecting median spine arising from basal half .... *Saica*

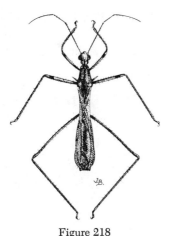

Figure 218

Figure 218 *Saica fuscovittata* Barber

This is an elongate slender species, 8-9 mm in length. It is nearly uniformly dull yellow in color, the femora having a subdistal, and the tibiae a subproximal, brownish ring. The front femora and tibiae are strongly curved. Usually some of the veins of the corium are tinged with crimson. It is reported only from the extreme southeastern states, although we have taken an individual at light in Connecticut. An additional species occurs in Texas.

4 Second antennal segment subdivided into a series of annular "pseudosegments" .... *Hammatocercus*

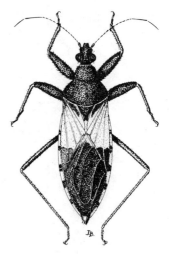

Figure 219

Figure 219 *Hammatocercus purcis* (Drury)

This is a large (22-30 mm) handsome insect with the head, pronotum and scutellum black, anterior portion of the hemelytra white and the abdominal connexivum red. The annulated second antennal segment is diagnostic. It lives under bark. The range is from Virginia and Indiana, south to Florida and Texas.

4a Second antennal segment terete, never subdivided into a series of annulate pseudosegments 5

5 Front and middle tibiae without a terminal plate, although frequently with a small hairy convex pad present 6

5a Front and usually middle tibiae ending in an oval or oblong lobe or plate which is concave and more or less spongy beneath (sometimes very small, i.e. *Triatoma*) 21

6 Distance between ocelli greater than distance between compound eyes .... *Apiomerus*

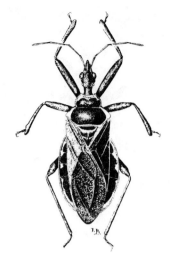

Figure 220

Figure 220 *Apiomerus crassipes* (F.)

This is a broadly oval thick-bodied medium-sized (14-19 mm) insect. It is chiefly black with the pronotum, abdomen and scutellum posteriorly having red to reddish-yellow or orange margins. The body surface is rather thickly clothed with prominent upright hairs. The distribution is throughout most of the United States.

A. spissipes (Say) is a closely related species that can be separated from *A. crassipes* by the bright red corium and pronotum. The posterior margin of the pronotum and scutellum and the apical corial margins are yellow. It ranges from Florida to Texas and Arizona and has also been recorded from Colorado. Seven additional species occur in the southern and southwestern states.

6a Distance between ocelli less than distance between compound eyes 7

7 First antennal segment slender, only slightly thicker than segment two (fig. 222); cell in fore wing immediately anterior to two elongate membrane cells usually quadrate 8

7a First antennal segment usually thickened, if relatively narrow then always conspicuously thicker than segment two (fig. 234); usually with a hexagonal cell in fore wing immediately anterior to the two elongate membrane cells 17

8 Front lobe of head twice as long as hind lobe .... *Rhynocoris*

Rhynocoris ventralis (Say)

This is a moderately small species (10-11 mm) with the head, antennae, disc of the hind lobe of the pronotum, scutellum and legs black. The front lobe of the pronotum, the margin of the hind lobe, the coxae and the abdomen are dull red. However, some varieties have orange to red legs, antennae and coria. The distribution is from New England across the northern and central states to California. There are several color varieties that may have geographic significance.

A second species has been reported from Sitka, Alaska.

8a Front lobe of head not conspicuously longer than hind lobe, or if so, not twice length of hind lobe 9

9 Sides of mesopleuron with a small tubercle or fold projecting in front of posterior margin of propleuron. Fig. 221 10

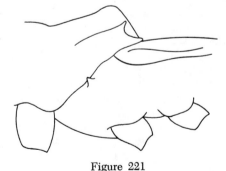

Figure 221

9a Sides of mesopleuron without a tubercle or fold 12

10   Pronotum produced caudad with a high median tuberculate ridge .... *Arilus*

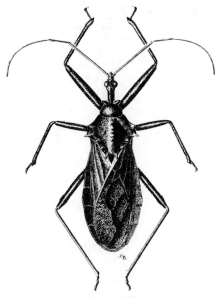

Figure 222

Figure 222   *Arilus cristatus* (L.)

This is one of the largest (28-36 mm) and most easily recognized of the assassin bugs. It is a uniformly dark brown color and the body surface is thickly covered with decumbent grayish pubescence. The antennae, labium, tibiae and tarsi are reddish brown. Its most distinctive feature is the bizarre semi-circular crest of the pronotum which bears eight to twelve thick heavy tubercles. It is often called the "wheel-bug" because of this feature. The range is from New York west to Illinois and south to Florida.

10a   Pronotum not produced caudad and without a high median ridge 11

11   Front tibiae unarmed .... *Acholla*

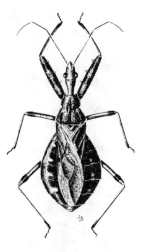

Figure 223

Figure 224

Figures 223 and 224   *Acholla multispinosa* (DeGeer)

This is a nearly uniformly dark brown to dull brownish-yellow medium-sized (12-16 mm) species. It is sparsely clothed with short grayish pubescence. The legs are vaguely annulate with brown and yellow. As with many assassin bugs the abdomen of the female (fig. 223) is much broader than is that of the male (fig. 224). The range is from New England west to Wisconsin and Colorado and southwest to Arizona. Two additional species occur in the southwestern states.

11a Front tibiae possessing prominent spines on inner surface .... *Sinea*

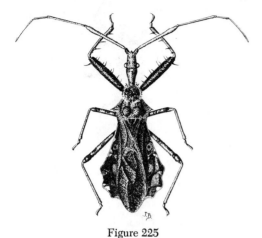

Figure 225

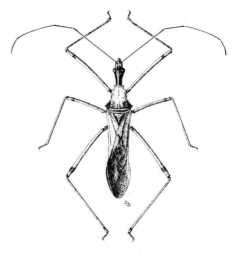

Figure 226

Figure 225 *Sinea diadema* (F.)

This is a nearly uniformly dark brown to dull reddish-brown species. It is from 12-14 mm long and is sparsely clothed with fine decumbent yellowish pubescence. The connexivum has a pale spot at the caudo-lateral angle of each segment. This is one of the most common of North American reduviids. It is frequently captured in weedy fields on the flowers of Compositae and other plants. The distribution is nearly throughout the United States.

Sinea spinipes (F.) closely resembles *diadema* but the anterior pronotal lobe has only blunt tubercles whereas in *diadema* distinct sharp spines are present in this area. This is a widely distributed species east of the Rocky Mountains. Eight or nine additional species occur in the southern and southwestern states.

12   First segment of labium scarcely or not longer than one-half length of second segment .... *Zelus*

Figure 226 *Zelus exsanguis* (Stål)

This is an elongate, slender insect (15-17 mm). In life it is a bright green color that fades to dull yellow in preserved specimens. The head, hind lobe of the pronotum and the scutellum are sometimes dark brown to black. Each humeral angle of the pronotum has a short acute dark spine. This is a very common arboreal species typically found on shrubs and trees along roadsides and in wooded areas. In the north it overwinters as a striking green nymph with golden abdominal markings. The range is from New England west to the Pacific and south to Florida and Texas. About a dozen additional species are known from the United States, most restricted in distribution to the southern and southwestern states.

12a First segment of labium much longer than one-half length of segment two .. 13

13   First segment of labium shorter than second segment .... *Pselliopus*

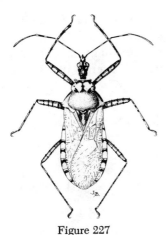

Figure 227

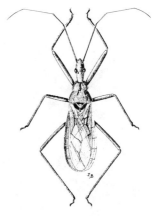

Figure 229

Figure 227 *Pselliopus cinctus* (F.)

This is a striking species of a dull orange color with the head and often the front lobe of the pronotum mark with black and white, and with the legs and the first two segments of the antennae with numerous black and white bands. The length is 12-13 mm. The range is from Massachusetts south to Florida and west to Wyoming and Texas. Five additional species are known from the United States.

13a First segment of labium as long as or longer than second segment 14

14 Hind lobe of pronotum without spines or with very short ones .... *Fitchia*

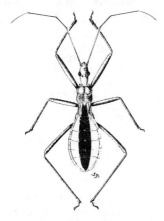

Figure 228

Figures 228 and 229 *Fitchia aptera* Stål

This is an elongate (12-14 mm) elliptical dull yellow species which is easily recognized by the presence of a broad dark longitudinal stripe on either side of the midline on the dorsal surface of the abdomen, and by a broad dark lateral stripe midway between the meson and the lateral margin of the abdominal sternum. It is without spines on the hind lobe of the pronotum. This species is generally micropterous with the wings reduced to tiny pads that barely extend onto the first abdominal segment (fig. 228). The macropterous form is known (fig. 229), but is rare. Generally this species occurs on the ground about grass clumps in old fields. It is widely distributed from Maine west to Utah and southwest to Texas, but is generally uncommon.

The only other United States species, *F. spinulosa* Stål, may be distinguished by having two short spines or tubercles on the hind lobe of the pronotum. It ranges from New York to Indiana and south to Florida and Texas.

14a Hind lobe of pronotum armed with four elongate, tapering prominent spines. Fig. 230 15

15 Caudo-lateral angles of abdominal sterna three to five prolonged posteriorly into distinct spines; only a short acute erect spine behind base of each antenna .... *Atrachelus*

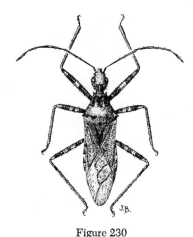

Figure 230

Figure 230 *Atrachelus cinereus* (F.)

This is one of the smallest (7-8 mm) of the United States reduviids. It is slender and brownish yellow in ground color but thickly covered with gray pubescence. It is readily recognizable by its small size and the prolonged spines on the abdominal connexivum. The range is from Pennsylvania west to Michigan, south to Florida and southwest to Texas.

15a   Caudo-lateral angles of abdominal sterna three to five not prolonged posteriorly into distinct spines; a very long tapering acute erect spine present behind base of each antenna 16

16   Fore femora distinctly incrassate; head spines pale, their length half interocular space; hemelytra wholly pale .... *Rocconota*

Rocconota annulicornis (Stål)

This is an elongate-oval rather stout species (16-20 mm). The color varies from reddish to straw-yellow and the body surface is thickly clothed with a fine yellow pubescence. The first antennal segment has two broad brown bands. It appears to be a rare species that occurs from New Jersey south to Florida and southwest to Texas.

16a   Fore femora not markedly incrassate; head spines black or very dark brown, nearly as long as interocular width; hemelytra in great part black .... *Repipta*

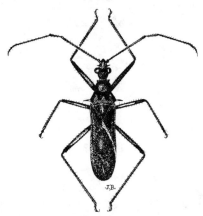

Figure 231

Figure 231 *Repipta taurus* (F.)

This is a slender species, 11-13 mm in length. It is easily recognizable by the predominately orange to red color with four contrasting black stripes on the hind lobe of the pronotum. The range is from Pennsylvania south to Florida and southwest to Texas.

Two additional species are known from the southwest.

17   Head with one or more branched spines or processes on each side below and posterior to eyes. Fig. 232 18

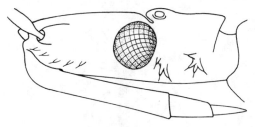

Figure 232

17a   Head lacking branched spines or processes below and behind eyes 19

18 Distal end of first antennal segment produced forward as a conspicuous blunt spine .... *Pnirontis*

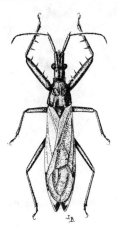

Figure 233

Figure 233 *Pnirontis infirma* Stål

This is an elongate slender (10-12 mm) bug of a uniformly dull yellow color with a prominent black to dark brown spot at the caudo-lateral angle of each abdominal connexivum. The range is from New Jersey west to Illinois and south to Florida and Texas.

Four additional species are found chiefly in the southeastern states.

18a Distal end of first antennal segment not produced into a spine .... *Pygolampis*

Figure 234

Figure 234 *Pygolampis pectoralis* (Say)

This is an elongate (13-16 mm) slender, parallel-sided dark reddish-brown species clothed with fine decumbent silvery pubescence. It frequently comes to light. The range is almost throughout the United States with the possible exception of the northwestern states.

P. sericea Stål is a closely related species that ranges from New England west to Illinois and southwest to South Carolina and Texas. It can be distinguished from *P. pectoralis* by its distinctly swollen front femora and by having the first antennal segment subequal in length to the ante-ocular part of the head. The front femora of *P. pectoralis* are only slightly swollen and the first antennal segment is nearly twice as long as the ante-ocular part of the head.

19 Large species, over 20 mm in length, hind portion of head not strongly narrowed into a "neck" (but neck-like in portion of head that is inserted into thorax). Fig. 235 .... *Stenopoda*

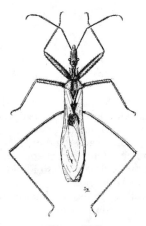

Figure 235

Figure 235 *Stenopoda spinulosa* Giacchi

This is a very large (22-30 mm) elongate, robust, nearly uniformly straw-yellow species. The third and fourth antennal segments, longitudinal stripes on the membrane and the quadrate cell immediately anterior to the membrane are contrasting dark chocolate-brown to black. The front femora are armed beneath with a row of very short tubercles in addition to rows of prominent elongate hairs. The range is from New Jersey south to Florida and west to Kansas and Texas.

19a Small species, 18 mm or less in length, hind portion of head strongly narrowed into a "neck". Fig. 236 20

20 Fore femora thickened, armed beneath with short stout spines ...... *Oncocephalus*

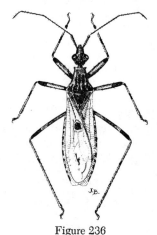

Figure 236

Figure 236 *Oncocephalus geniculatus* Stål

This is an elongate-oval pale yellowish-brown species with the hind lobe of the head, the disc of the hind lobe of the pronotum, two large patches on the wing membrane and the scutellum dark brown. Females are apparently brachypterous. The range is from North Carolina south to Florida and west to Colorado and Texas.

Two other species are known from the central and southwestern states.

20a Front femora but little thickened, unarmed beneath .... *Narvesus*

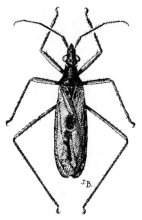

Figure 237

Figure 237 *Narvesus carolinensis* Stål

This is an elongate, parallel-sided species, 14-17 mm in length. It is dull straw-yellow in color with scattered brownish markings. It closely resembles *Oncocephalus geniculatus* in habitus. The range is from New Jersey south to Florida and west to Missouri and Arizona.

21 Apex of scutellum broad, with two (or three) large heavy curving spines present .... *Rhiginia*

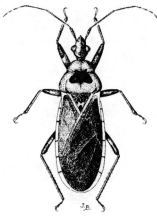

Figure 238

128   Families of Hemiptera

Figure 238 *Rhiginia cruciata* (Say)

This is an oblong-oval species of moderate size (12-16 mm). It is of a contrasting black and reddish to reddish-yellow color with the black coloration largely confined to the scutellum, a spot or median patch on the disc of the hind lobe of the pronotum, the sterna and the hemelytra except for the base of the corium. The legs are yellow with the distal one-third to one-half of the femora and the tibiae blackish, except for a broad submedian pale band. The range is from New Jersey west to Illinois and south to Florida and Texas.

A second species is known from Texas and New Mexico.

21a Apex of scutellum narrow, without spines, or prolonged posteriorly as a single spine 22

22 Front coxae with outer side flat or concave 23

22a Front coxae terete, their outer sides convex 25

23 Middle tibiae lacking a spongy fossa .... *Sirthenea*

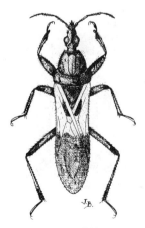

Figure 239

Figure 239 *Sirthenea carinata* (F.)

This is a very large (18-25 mm) elongate insect. The head, pronotum, scutellum and under surface are black or very dark brown with at least the basal half of the clavus and the greater part of the corium a strongly contrasting orange-red. Little is known of its habits. It ranges from New Jersey and Ohio south to Florida and west to Kansas, Texas and California.

23a Middle tibiae with a spongy fossa present on under side 24

24 Distal fourth of front tibiae angularly dilated in front of the spongy fossa; hemelytra wholly black or reddish black .... *Melanolestes*

Figure 240

Figure 240 *Melanolestes picipes* (H. S.)

This is a moderate-sized (15-20 mm) robust black species having a general habitus resemblance to *Reduvius personatus* (L.). The hemelytra of the females are often reduced to small pads. This is a common species that ranges from New England west to the Great Plains and south and southwest to Florida, Texas and California.

M. abdominalis (H.-S.) has a red abdomen and perhaps is only a color variety of *picipes*. It ranges from New England west to South Dakota, south to Mississippi, Texas and California.

Families of Hemiptera

24a Apical fourth of front tibiae but little (and not angulately) dilated; corium in part yellow, membrane with a large median yellow spot .... *Rasahus*

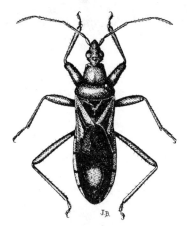

Figure 241

Figure 242

Figure 241 *Rasahus biguttatus* (Say)

This is a large (16-20 mm) species, predominantly dark reddish brown with the inner basal half of the corium and most of the clavus yellow, and a large oval yellow spot in the middle of the membrane. The legs are yellowish brown. It is a ground-living species, frequently taken under stones and it often comes to light. The distribution is southern from North Carolina south to Florida and west to Iowa, Texas and California.

Two additional species are known from Florida and the southwestern states.

25 Antennae inserted in lateral or dorso-lateral margins of head; antenniferous tubercles projecting slightly from sides of head .... *Triatoma*

Figure 242 *Triatoma sanguisuga* (Leconte)

This is a moderate sized (16-21 mm) species of a variegated reddish and dark brown to black color. The basal fourth and a spot on the apical third of the corium are reddish yellow. The connexivum is broad and marked with alternating red and black patches. This is an infamous insect known in the southern states as the "big bed bug." It is a nocturnal feeder, remaining hidden during the day. It often occurs in houses where it feeds on bed bugs and other insects as well as mammals, and it often bites humans. Its bite is severe and causes swelling accompanied by itching. It has been implicated in the transmission of Chagas fever. The range is from New Jersey west to Illinois and Kansas and south to Florida and Texas.

T. heidemanni Neiva is a closely related species having about the same range. It can be distinguished from T. *sanguisuga* by the long inclined hairs on the labium. About ten additional species occur in the United States, nearly all confined to Texas, Arizona or California.

25a Antennae inserted on top of head between margins close to eyes; antenniferous tubercles not projecting from sides of head .... *Reduvius*

Figure 243

Figure 243 *Reduvius personatus* (L.)

This is a nearly uniformly dark brown to almost black elongate-oval bug. It ranges from 17-22 mm in length. The head, pronotum and scutellum are shining or subshining. The body and appendages are clothed with prominent upright hairs. This species is often called the "masked bed bug hunter." It often occurs and breeds in houses where it feeds on various small insects including bed bugs. The nymphs cover their bodies with a thick coating of dust and look like small animated dust bundles.

R. personatus is a Holarctic species that ranges in North America from New England south to Florida and west to Kansas.

A second species is found in Arizona.

26   Winged 27

26a Apterous or micropterous 36

27   Vein M terminating basally on r-m cross vein of fore wing, not on submarginal vein (in difficult cases it can be seen that M terminates about as far out on wing as end of vein 1A (fig. 244); claws of all legs simple .... *Ploiaria*

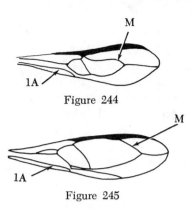

Figure 244

Figure 245

Ploiaria carolina (H. S.)

This is a small (5-6 mm) slender bug. It is dark brown in color with a pale median line on the head and pronotum. The fore femora have a pale ring at the distal third and the distal ends of the hind femora are yellowish. It is found from North Carolina to Florida.

P. denticauda M. & M., known from California, Arizona and Texas, and *P. hirticornis* (Banks) from North Carolina, District of Columbia, Louisiana and Florida are of the same general color and size as *P. carolina* but can readily be distinguished from it by both having a prominent median backward projecting spine on the posterior lobe of the head. *P. similis* M. & M. from Texas and *P. reticulata* (Baker) from Texas and California are both larger (8-9 mm) species. The two can be separated by the presence of a small erect spine on the margin of each eye on the posterior lobe of the head of *P. reticulata*. Eight additional species occur in the southern and southwestern states.

27a Vein M terminating on submarginal vein (R) far out from r-m (the termination of M basally is much further out on wing than distal end of vein 1A) (fig. 245); undersurface of claws with incisions or projections present 28

28   Fore femora with most basal spine of series of spines running along posteroventral surface conspicuously larger than preceding spines; fore tarsi usually unsegmented (three immovable segments in *Emesaya*); scutellum and metanotum without spines 29

28a Fore femora not having basal spine on postero-ventral series conspicuously larger than preceding spines; fore tarsi with movable segments; scutellum or metanotum often with spines 32

29 Fore tarsi three-segmented ...... *Emesaya*

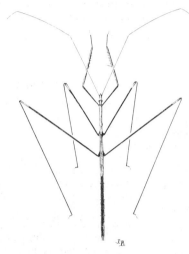

Figure 246

Figure 246 *Emesaya brevipennis* (Say)

This is the largest and most conspicuous of the thread-legged bugs. It is an elongate subcylindrical insect, 33-37 mm in length, and covered with a silvery pubescence. It is often found in barns, on porches or in old buildings. Specimens have been observed waiting in spider webs for other insects to become entangled upon which they then feed. The range is throughout most of the United States. Four additional species occur in the southern and southwestern states.

29a Fore tarsi unsegmented 30

30 Hind lobe of pronotum not prominently developed, leaving most of mesonotum exposed .... *Ghinallelia*

Ghinallelia productilis (Barber)

This is an elongate slender brownish insect covered with short, fine, yellowish pubescence. There is a distinct black dot on the upper surface of each fore femur near the distal end and a pair of dots in the middle on the hind lobe of the head and pronotum. The length is from 22-25 mm. In this country it is known only from Florida.

30a Hind lobe of pronotum well developed, completely covering mesonotum 31

Figure 247

31 Claws of middle and hind legs with a submedian projection. Fig. 247 .... *Pseudometapterus*

Pseudometapterus butleri Wygodzinsky

This is a very slender micropterous insect which is straw-yellow to brown in general color with a black lateral stripe on the head and sides of the thorax. It is known only from Arizona. Two additional species occur in Florida and in Texas.

31a Claws of middle and hind legs simple, lacking a submedian projection .... *Barce*

Barce fraterna (Say)

This species is dull yellow in general color above with the ventral surface dark brown. The disc of the hemelytra has numerous small dark spots. The length is 13-15 mm. It ranges from New York west to Oklahoma and Kansas and south to Florida and Texas.

  B. uhleri (Banks) (fig. 248) is smaller (7-10 mm) than *B. fraterna* and dark brown in general color. It is restricted to the eastern United States. Four additional species occur in the southern states.

Figure 248

Figure 249

32 Ventral surface of fore tibiae with only well developed setae, lacking strongly sclerotized spinulets or denticles; usually small insects, well under 10 mm .... *Empicoris*

Empicoris errabundus (Say)

This is a small (4-5.5 mm) slender elongate species. The head and the front lobe of the pronotum are brown with the posterior pronotal lobe yellowish. The first segment of the antennae is brown with 10-12 narrow rings and with the apex pale yellow. Antennal segments two and three are dull yellow with the proximal half of segment two having three or four vague brownish rings. The hind wings have conspicuous dark spots apically. This is the common and most widely distributed species in the genus. It is known from New England south to Virginia and southwest to Kansas and Texas. It has also been recorded from Oregon and Washington.

    E. orthoneuron M. & M. is a rather variable species from 4-6 mm in length with the color pattern lacking any uniformity. It can be differentiated from *E. errabundus* by the lack of dark spots on the apical portion of the hind wings. It is known from Oregon, California, Nevada and Texas.

    E. rubromaculatus (Blatchley) is also widely though spottily distributed, being recorded from California, Indiana, Mississippi, Virginia and Florida. It can readily be distinguished from other species in the genus by the lateral carinae of the hind lobe of the pronotum being distinguishable only at the anterior extremities whereas the lateral carinae are complete in other species. Ten additional species occur north of Mexico.

32a Ventral surface of fore tibiae with strongly sclerotized spinulets or denticles in addition to setae (fig. 249); larger, usually more than 10 mm 33

33 Fore wing with a single closed cell .... *Gardena*

Gardena elkinsi Wygodzinsky

This is a reddish-brown species without any distinct markings. The hemelytra are dark hyaline. The eyes are large. The length is from 9-11 mm. It is known from California, Arizona, New Mexico and Florida.

    G. poppaea M. & M., the only other species known from the United States, is found in Texas and Florida and is nearly twice as long (17-20 mm).

33a Fore wing with one or two smaller cells in addition to large discal cell 34

34 Fore tarsi three segmented .... *Emesa*

Only one species *annulata* (Dohrn) is known to occur in the United States (Arizona). All other members of this genus are Neotropical in distribution.

34a Fore tarsi two-segmented 35

35 Very hairy insect, especially on legs; hind lobe of pronotum with a pair of short projections on dorsal surface .... *Stenolemus*

Stenolemus lanipes Wygodzinsky

This is a curious looking insect with a short wide head, very large eyes, a slender "neck" and with the lobes of the prothorax formed to resemble a dumbbell. The length is 9-11 mm. It is pale clay-yellow. It is known from the southeastern United States. Three additional species occur in the southern and southwestern states.

35a Not conspicuously hairy except perhaps on hind tibiae and on antennae; hind lobe of pronotum lacking dorsal projections .. .. *Stenolemoides*

Stenolemoides arizonensis (Banks)

This is a pale brownish-yellow species without distinct markings on the wings. The female is brachypterous. The proximal two segments of the antennae have white rings. The anterior femora and tibiae also have faint white rings. The length is 8-9 mm. The range includes California, Nevada, New Mexico and Texas.

36 Fore femora not having basal spine on postero-ventral series conspicuously larger than preceding spines; fore tarsi with movable segments; scutellum or metanotum often with spines .... *Gardena* (see 33)

36a Fore femora with most basal spine of series of spines running along postero-ventral surface conspicuously larger than preceding spines; fore tarsi usually unsegmented; scutellum and metanotum without spines 37

37 Fore tarsi two-segmented, subequal in size to middle and hind tarsi .... *Empicoris* (see 32)

37a Fore tarsi either unsegmented or three-segmented, conspicuously longer than middle and hind tarsi 38

38 Fore tibiae with two irregular series of conspicuous spiniferous processes; fore tarsi not bare above and at sides 29

38a Fore tibiae lacking conspicuous spiniferous processes; fore tarsi bare above and at sides .... *Ploiaria* (see 27)

REFERENCES

Readio, P. A. 1927. Studies on the biology of the Reduviidae of America north of Mexico. Univ. Kansas Sci. Bull. 17:(1):5-291.

Wygodzinsky, P. W. 1966. A monograph of the Emesinae (Reduviidae, Hemiptera) Bull. Amer. Mus. Nat. Hist. 133:1-614.

NABIDAE
The Damsel Bugs

The majority of nabids are dull brown in color, relatively slender, with a narrow elongate head, a four-segmented beak and very slender antennae. The pronotum is two-lobed. It is either singularly appropriate or inappropriate, depending upon one's point of view, that the members of such a fiercely predatory family should be known as damsel bugs. All species feed upon other insects and search actively for their prey on shrubs, weeds and on the ground. A remarkable genus occurring in the West Indies lives in spider webs. The nymphs of some species are strongly ant mimetic.

1 Antennae five-segmented; clavus not distinctly widened posteriorly .... *Pagasa*

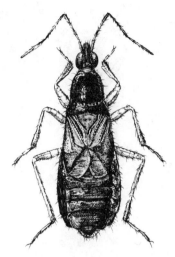

Figure 250

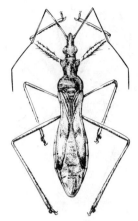

Figure 251

Figure 250 *Pagasa fusca* (Stein)

This is a short, thick-bodied damsel bug which is easily recognized by its shining black or dark brown color which contrasts strongly with the yellow or orange-yellow legs and labium. It is a ground-living insect, usually found running about in dry hot areas among grass and weeds, searching for prey. It is not uncommon if ground collecting is done extensively. Both short- and long-winged forms occur. The range is throughout the United States but it is scarce in the southern states where its distribution is poorly understood. Two additional species occur in the southern and southwestern states.

1a   Antennae four-segmented; clavus widening posteriorly 2

2   Anterior coxae slender and greatly elongated; ocelli absent; wings strongly constricted before the middle. Fig. 251 .... *Carthasis*

Figure 251 *Carthasis decoratus* (Uhler)

This is a small (3.5-5 mm) elongate slender insect which is easily distinguished from most other damsel bugs by the strong constriction near the middle of the body. The coloration is yellow or greenish yellow, and frequently there are red or brown markings prominently scattered over the body. The species is usually found on low trees and bushes. It is a scarce species ranging from New York south through the eastern states to Florida and west to Mississippi.

2a   Anterior coxae not greatly elongated; ocelli present; wings usually not strongly constricted before the middle 3

3   First antennal segment at least twice as long as head, membrane lacking closed cells .... *Metatropiphorus*

Figure 252

Figure 252 *Metatropiphorus belfragei* Reuter

This is a small (6-7 mm) very slender and elongate insect which is dark brown in color. It lacks spongy pads (fossa) on the ends of the front and middle tibiae. The body is thickly covered with grayish pubescence. It is a scarce species found on trees and tall shrubs. The range is through the eastern and central states from Florida and Texas north to Iowa and Connecticut.

3a First antennal segment less than twice length of head; membrane of fore wing in macropters usually with closed cells 4

4 Body shining black, legs and lateral margins of abdominal connexiva yellow .... *Nabicula*

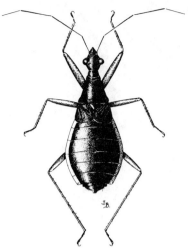

Figure 253

Figure 253 *Nabicula subcoleoptrata* (Kirby)

This is a shining black insect with the legs, antennae, labium and lateral connexival margins of the abdomen bright yellow. The nymphs are remarkable ant mimics, resembling large black ants in appearance and movement. It is found in meadows where it feeds chiefly on the meadow plant bug, *Leptopterna dolabrata* (Linn.). Macropterous forms are extremely rare; no long-winged males are known. Its distribution is northern across the continent from the east coast to British Columbia and Colorado. It is not known to occur south of New Jersey and Kansas.

4a Body chiefly or entirely brown or gray in color 5

5 Tibiae annulate for entire length; fore and middle femora (fig. 254) armed below with small, short black teeth .... *Hoplistoscelis*

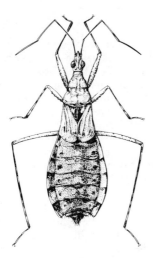

Figure 254

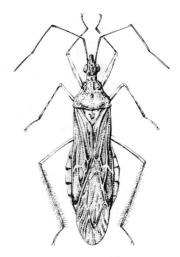

Figure 255

Figure 254 *Hoplistoscelis sordidus* (Reuter)

This species is easily recognizable by the annulate (ringed) tibiae and femora, and especially by the close set series of tiny black teeth present on the underside of the front and middle tibiae. It varies in color from mottled dark brown to a very light yellow. The short-winged form is generally much more common than the long-winged. It is found in shady moist areas with luxuriant vegetation. It ranges over the entire eastern United States west to the Great Plains. Four additional species occur in the southern and western states.

5a Tibiae usually not annulate, or if so then annulations confined to proximal and distal ends of tibiae; fore and middle femora unarmed, or with thin black spinelike setae, never bearing blunt teeth 6

6 Posterior pronotal lobe strongly and thickly punctate; hind tibiae clothed with very elongate erect hairs; hemelytra shallowly but distinctly constricted before middle .... *Lasiomerus*

Figure 255 *Lasiomerus annulatus* (Reuter)

This is a large pale yellowish species obscurely mottled with brown. The apices of the coria and the annulations near the distal ends of the femora and proximal ends of the tibiae are contrastingly dark brown, or sometimes crimson. It is usually found on rank undergrowth in woodlands, and it flies readily. The range is over the eastern United States west to Iowa and Missouri. Two additional species occur in the southern states.

6a Posterior pronotal lobe either completely impunctate or at most with a few scattered punctures; hind tibiae clothed with relatively short hairs that project at a sharp angle to tibial surface; hemelytra of macropters with costal margins nearly parallel, not distinctly constricted before middle 7

7 Scutellum with a prominent, conspicuously depressed, shining semicircular spot at each antero-lateral angle .... *Dolichonabis*

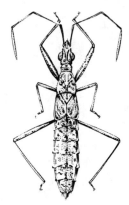

Figure 256

Figure 256 *Dolichonabis propinquus* (Reuter)

This is the most elongate of the North American damsel bugs (11-12 mm). It is easily recognized by the elongate body and by the shining basal scutellar spots. Macropterous individuals are very rare (male macropters unknown). Brachypters usually have a dark brown or red stripe running the length of the abdomen along the midline. It lives on sedges and grasses in marshy areas. The distribution is northern from Ontario west to Alberta and south to Ohio, Maryland and Iowa. Two additional species are known from the northern states west to Colorado.

7a Scutellum without antero-lateral shining spots, or with spots obsoletely developed .... *Nabis*

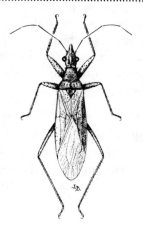

Figure 257

Figure 257 *Nabis americoferus* Carayon

This is a small dull gray-yellow insect (6-9 mm) usually with an obscure median dark stripe on the head and front lobe of the pronotum. The hind tibiae are uniformly pale. This is the most common nabid and one of the most common hemipterans throughout the United States. It lives in drier places than do most of the related species, often occurring in gardens and fields. The eggs are laid in grass stems. Most records in the North American literature refer to this species under the name *Nabis ferus* (L.).

N. alternatus Parshley is very similar to *americoferus* and often confused with it, but it usually has dark red-brown spots on the hind tibiae and usually has a longer first antennal segment that is equal to the width of the head across the eyes. The wings are markedly speckled with brown. It is distributed over the entire western part of the country east to the Mississippi River.

N. roseipennis Reuter is a yellow to red-brown insect, usually extensively marked with irregular dark lines and spots. There is usually a large black spot behind each eye. It is one of our larger species (7-9 mm), is usually darker brown than most other closely related species and has strongly spotted tibiae. Both long and short winged forms occur. It is found chiefly along wooded margins or in the shady parts of fields. The species ranges everywhere east of the Mississippi River and in the northwest to British Columbia, Alberta and Colorado.

Approximately ten additional species are present north of Mexico.

REFERENCES

Harris, Halbert M. 1928. A monographic study of the hemipterous family Nabidae as it occurs in North America. Ent. Amer. IX (1 and 2):1-97.

POLYCTENIDAE
The Bat Bugs

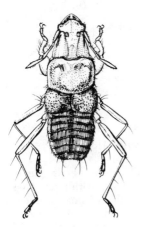

Figure 258

These rare insects are ectoparasites of bats. They are flightless and lack both compound eyes and ocelli. The head and prothorax are both greatly flattened. The anterior legs are short and flattened while the middle and hind legs are long and slender. The body is generally covered with short bristles and some longer bristles are also present on the legs. They range in size from 3.5-4.5 mm. Two species occur in the United States—*Hesperoctenes eumops* Ferris and Usinger which is found only in California on the free-tailed bat (*Eumops californicus*), and *Hesperoctenes hermsi* Ferris and Usinger which is known only from Texas and is found on the free-tailed bat (*Tarida macrotis*). A closely related species from Guatemala, *Hesperoctenes longiceps* (Waterhouse), is illustrated (Fig. 258).

REFERENCES
Ferris, G. F. & R. L. Usinger. 1939. The family Polyctenidae (Hemiptera: Heteroptera). Microentomology 4:(1):1-50.
Usinger, R. L. 1946. General catalogue of the Hemiptera Fasc. V. Polyctenidae. Smith College, Northampton, Mass. 18 pp.

CIMICIDAE
The Bedbugs

This is a small family all of whose members are bloodsucking ectoparasites of mammals and birds. The family takes its common name from the "human" bed bug which feeds primarily on man. Bedbugs are strongly flattened dorso-ventrally and are often reddish brown in color. The front wings are always reduced to small pads and there is no trace of the hind wings. All species live away from the host animal except when feeding, usually living in cracks, crevices or in nests.

Most of these highly specialized insects have a most unusual manner of insemination. The male does not copulate in the usual way but punctures the body wall of the female to inject the spermatozoa into a special organ—the spermalege. From here the spermatozoa migrate through the body cavity to the female genital tract. This interesting phenomenon with which are associated many fascinating evolutionary questions is sometimes aptly enough called "traumatic insemination."

The distributions of many of our bedbugs are poorly known, and the general collector can add significantly to our knowledge by making a special effort to obtain these cryptic ectoparasites.

1   Tibiae mottled; apices of tarsi with several stout spines on opposite side from insertion of claws. Fig. 259 .... *Primicimex*

Figure 259

Primicimex cavernis Barber

This is a very primitive monotypic bedbug over 7 mm long with very long legs, a heavy body, and an elongate head. It is the only known genus that lacks a spermalege. This remarkable insect was described from Ney Cave near Bandera, Texas where it probably lived on the Mexican free-tailed bat (*Tarida mexicana*). This colony is apparently destroyed. The only other known colony is in Guatemala.

1a Tibiae not mottled; tarsi lacking apical spines in opposition to claws 2

Figure 260

2   Bristles at lateral sides of pronotum with small teeth on outer sides (fig. 260); metasternum forming a flat plate between coxae 3

2a Bristles at lateral sides of pronotum lacking small teeth on outer sides; metasternum a somewhat compressed rounded lobe between coxae 4

3   Body clothed with pale slender bristles, second antennal segment not more than 2/3 length of interocular space; pronotum less than 1.5 times head width. Fig. 261 .... *Oeciacus*

Figure 261

Oeciacus vicarius Horvath

This is a small, elliptical bedbug covered with long, silky, pale hairs and with relatively short and thick third and fourth antennal segments. It is found chiefly in the nests of cliff swallows although it also occurs occasionally in those of barn swallows. The "swallow bug" is widely distributed in North America but apparently relatively scarce in the northeastern states and as yet unknown in the southeast.

3a Body clothed with relatively short stout bristles; second antennal segment about equal in length to interocular space; pronotum at least 1.5 times as wide as head. Fig. 262 .... *Cimex*

Figure 262

The members of this genus are the "true" bedbugs and include parasites of man. There are eight species found in the United States, all but the domestic species being found on bats. The most common and widespread species may be separated by the following key.

a   Females with hind margin of fourth visible abdominal sternum narrowly cleft or grooved (paragenital sinus) on right side (fig. 263); hind femora usually more

140   Families of Hemiptera

than 2.6 times as long as wide ....
.... *Cimex lectularius* L. (Fig. 264)

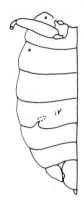

Figure 263

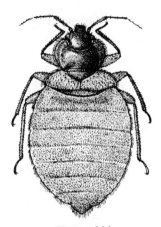

Figure 264

This is the common bedbug of man that gives the family its common name. It is found over nearly the whole world. These flat reddish-brown insects live in concealed places and generally crawl upon people when they are sleeping and suck their blood. It is believed that the ancestor of this pest originally was a bat parasite in the old world and subsequent to adopting man as a host has spread with him to all continents. The bedbug, while now primarily a parasite of man, also commonly attacks chickens and bats and has been reported from many other mammals and birds.

The tropical human bedbug *Cimex hemipterus* (F.) has been reported from Florida. It may be distinguished from *lectularius* by having the pronotal width less than 2.5 times the median length, whereas in *lectularius* the width is greater than 2.5 times the median length.

aa Paragenital sinus on right side of fourth visible abdominal sternum roundly emarginate (fig. 265); hind femora usually less than 2.6 times as long as maximum width .... b

b Longest bristles of hind tibiae 0.8 or less times as long as width of tibia; long lateral pronotal bristles less than 0.2 mm and distinctly serrate (western United States) .... *Cimex pilosellus* Horvath

Figure 265

This is a western species which is found on bats in the Far West from California to British Columbia and east to Montana in the north. It has been reported as attacking many bats of the family Vespertilionidae. Three very closely related species also occur on bats in the west.

bb Longest bristles of hind tibiae almost as long as width of tibiae; lateral pronotal bristles over 0.2 mm and very feebly or not at all serrate ....
.... *Cimex adjunctus* Barber

This species is closely related to the Western Bat Bug (*C. pilosellus*) but has much longer lateral pronotal bristles. It occurs as a parasite of bats throughout eastern and central North America, west to Texas and Colorado. A closely related species *C. brevis* is known from Illinois, Michigan, Minnesota and Quebec.

4   Bristles at sides of pronotum numerous and of equal length, lacking one or two

exceptionally long bristles at postero-lateral angles .... *Hesperocimex*

Hesperocimex coloradensis List

This species is a parasite of purple martins (*Progne subis*) and woodpeckers and is usually found in abandoned woodpecker holes. It is easily recognizable by the color pattern. The entire central area of the abdominal sternum is clear and membranous but laterally darker colored. It is known from California, Colorado, Oregon and British Columbia. Curiously it has never been taken in man-made martin houses in the Central and eastern United States. *H. sonoriensis* has been taken in Arizona and Baja California.

4a Bristles at sides of pronotum short and variable but always with one or two long bristles at postero-lateral pronotal angles. Fig. 266 5

Figure 266

5 Head as wide as long, gula swollen at arched front margin of prosternum .... *Synxenodorus*

Synxenodorus comosus List

This species is recognizable by the possession of 2-3 very long bristles along the lateral pronotal margin anterior to the postero-lateral ones and by the swollen ventral head surface. It is known only from Nebraska and California in nests of the white-throated swift (*Aeronautes saxatalis*).

5a Head wider than long; gula not swollen 6

6 Labium relatively elongate, reaching beyond apices of middle coxae; second antennal segment subequal to interocular space .... *Haematosiphon*

Haematosiphon inodorus (Duges)

This species is easily recognized by the long labium. It is parasitic on owls, eagles, the California condor and chickens. It occurs in the southwestern United States from Oklahoma and Texas to California.

6a Labium shorter, scarcely reaching beyond the apices of the front coxae; second antennal segment shorter than interocular space 7

7 Pronotum approximately twice as wide as long, greatest width in front of middle .... *Cimexopsis*

Cimexopsis nyctalis List

This is a small, elliptical species, found over the entire eastern United States from Maine to Florida and west to Nebraska and Arkansas. It is a parasite of the chimney swift (*Chaetura pelagica*).

7a Pronotum at least 2 1/2 times as wide as long, greatest width behind middle .... *Ornithocoris*

Ornithocoris pallidus Usinger

This species is recognizable by the broad pronotum, the very short bristles and oval wing pads. It is considered to be a native of South America and has probably been introduced into the southeastern United States. It is known from Georgia and Florida and is parasitic on purple martins and chickens.

REFERENCE
Usinger, R. L., 1966. Monograph of Cimicidae (Hemiptera-Heteroptera). Thomas Say Foundation 7:VI-XI: 1-585.

ANTHOCORIDAE
The Minute Pirate Bugs

The members of this family are very small insects from 2 mm to at most 5 mm in length. They have flattened bodies with somewhat pointed heads, ocelli present, an apparently three-segmented labium and a distinct cuneus present in the front wing. The male genital capsule is asymmetrical and many species have the peculiar ectodermal or "traumatic" method of insemination found in the Cimicidae.

Anthocorids are chiefly predaceous, feeding upon a variety of small arthropods and some are of distinct economic benefit. A few are phytophagous. Some species occur on flowers, others on coniferous and deciduous trees, under bark, in grain bins and in the nests of birds and mammals. Despite their predaceous habits a number of species appear to occur only on a restricted number of plants.

More than twenty genera and seventy species have been reported from North America north of Mexico. The distribution of most species is very poorly understood and many are apparently restricted to the southwestern and northwestern states.

1   Fore femora armed below with several short stout teeth .... *Scoloposcelis*

Scoloposcelis flavicornis Reuter

This is a moderate-sized (2.8-3.5 mm) dark brown to blackish species with brownish-yellow wings that are darkened along the lateral margins. The head is elongate and the posterior pronotal margin is deeply concave. Species of this genus are often collected with bark and fungus beetles. It is known from such widely separated states as Pennsylvania, Florida, Texas, Indiana, and Idaho.

S. mississippiensis D. & H. differs from *flavicornis* in having the anterior and posterior femora equally incrassate, shorter hemelytra, the pronotum broader anteriorly and a longer labium that extends posteriorly to the hind coxae. It occurs in the Gulf states.

S. occidentalis D. & H. has a shorter labium that extends posteriorly only to the middle coxae, and the metathoracic scent gland channel is more strongly produced. It occurs in California. Two additional species occur in the Southwest.

1a   Fore femora unarmed below 2

2   Third and fourth antennal segments generally much more slender than preceding segments (fig. 267) and always having numerous elongate hairs which are much longer than diameter of a segment 3

2a   Third and fourth antennal segments short, rounded, nearly as thick as basal segments (fig. 273); hairs may be prominent but are always short 6

3   Lateral pronotal margins narrowly but distinctly carinate and reflexed; end of abdomen lacking elongate hairs .... *Lyctocoris*

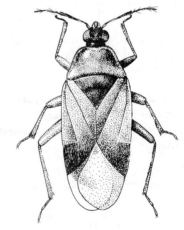

Figure 267

Figure 267 *Lyctocoris campestris* (F.)

This is a relatively large (3.3-3.9 mm) reddish or reddish-brown species. The membrane of the front wings is a dusky, transparent brown color with only a single distinct vein present. *L. campestris* is a semi-domestic species and is usually found in granaries, houses, haystacks, and bird and mammal nests. It has frequently been reported biting man. It ranges widely almost throughout North America.

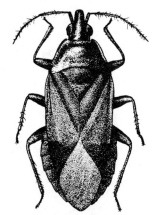

Figure 268

    L. stali (Reut.) (fig. 268) is a similar reddish-brown species which may be distinguished from *campestris* by having four distinct veins in the membrane of the fore wing and by having a longer labium which reaches the hind coxae. The distribution is across the country from New York to California.

    Seven additional species occur north of Mexico chiefly in the Pacific Northwest. Several of these are found almost exclusively on pines.

3a   Lateral pronotal margins rounded, or if carinate then not reflexed; end of abdomen with several very elongate hairs present 4

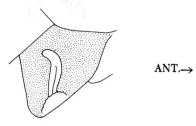

Figure 269

4   Scent gland channel curved backward (fig. 269); lateral margins of pronotum and corium bearing ciliate hairs .... *Lasiochilus*

Figure 270

Figure 270 *Lasiochilus fusculus* (Reut.)

This is a small species (2.5-3.0 mm) generally of a nearly uniformly dark brown color, with an almost impunctate clavus, a dusky membrane with a pale tip and with a fringe of very short hairs along the margins of the pronotum and fore wing. The distribution is apparently throughout the entire eastern United States, but it appears to be a rare and local species.

    L. pallidulus Reuter is a uniformly pale yellowish-brown species with four to five distinct rows of small punctures on the clavus. It is confined to Florida and the Gulf coastal states. In southern Florida, at least, it is an abundant and conspicuous species.

    Five additional species have been described from the southern states.

Figure 271

4a Scent gland channel curved forward (fig. 271); margins of pronotum and corium lacking a distinct series of ciliate hairs .. 5

5 Fore femora somewhat swollen; pronotal collar absent or at most with an indistinct transverse impression anteriorly on pronotum between deflexed anterior angles .... *Xylocoris*

Xylocoris cursitans (Fallen)

This is a small (2.2-2.5 mm) very dark brown shining species with light yellowish to dark brown wings. However, it is usually found in the brachypterous condition in which the wing covers extend only to the third abdominal segment. Fully winged females are known, although they are very rare.

X. cursitans is most frequently found under the bark of fallen, rotting logs where it presumably feeds on small springtails and mites. It has also been collected in granaries. The species is thought to be introduced from Europe and this seems likely as the known North American distribution is very spotty, with records from a number of northeastern states, the Midwest and the Pacific Northwest.

Xylocoris galactinus (Fieb.) is a somewhat larger species with very light straw yellow front wings that are often somewhat darkened at the tip of the corium. It is usually found in stored grain or straw. It has been reported from many eastern and central states from New York to Florida and also in the Pacific Northwest.

Xylocoris sordidus (Reut.) is similar to *galactinus* with a dull white or gray color, but *sordidus* is recognizable by the lighter color of the clavus and cuneus which have at most only the tips fuscous, and by the evenly curved middle of the scent gland channel. In *galactinus* the channel is distinctly angulate. This species is apparently confined to the southern states, but there are questionable northern records.

Four additional species are known, three from the Far West and one from New York.

5a Fore femora slender and unswollen; pronotal collar present, forming a distinct transverse impression before anterior angles .... *Calliodis*

Calliodis temnostethoides (Reuter)

This is a very small species (2.2-2.5 mm) with yellowish or dull white wings that become black on the cuneus and the apex of the corium. The head, pronotum and scutellum are dark, rich brown to black. The femora are black except at the distal end. Little appears to be known of its biology but in Connecticut it has been taken in slime fluxes. It is known from Connecticut, Illinois, New York, Texas and Florida, but is apparently a scarce species.

One additional species occurs in Florida.

6 Anterior pronotal collar poorly defined, placed entirely between deflexed anterolateral pronotal margins; scent gland channel very elongate, extending outward and forward to anterior margin of metapleuron (fig. 272); second antennal segment not longer than width of head across eyes .... *Orius*

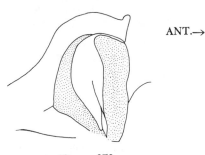

Figure 272

Figure 273

Families of Hemiptera   145

Figure 273 *Orius insidiosus* (Say)

This is a very small (2 mm) species with light yellowish wings that strongly contrast with the dark head, pronotum and scutellum. The distal area of the corium, the entire cuneus and antennal segment one are black or very dark brown. It is very frequently swept from various flowers and is an important predator of a number of crop pests such as the corn earworm. This is by far the commonest species of pirate bug in the entire eastern and central part of North America.

O. tristicolor (White) is very similar to *insidiosus* and replaces it ecologically in the western states. It is most readily distinguished by the completely dark clavus which is reliable for nine out of ten specimens, but genitalic differences are definitive.

Five additional species occur in North America, three on the Pacific coast, one in Florida and one in Iowa.

6a Anterior pronotal collar distinct, placed entirely or in part beyond antero-lateral pronotal margins; scent gland opening not reaching anterior margin of metapleuron (fig. 271), sometimes with a slender carina extending forward to margin; second antennal segment longer than width of head across eyes 7

7 Sides of scent gland channel elevated toward outer end; lateral pronotal margins distinctly flattened anteriorly; corium thickly and finely punctate .... *Tetraphleps*

Tetraphleps canadensis Provancher

This is a moderate-sized (3-3.9 mm), blackish-brown shining species with the wings pale brown but darker near the apices of the corium and cuneus. The membrane is semi-transparent brown with pale basal and lateral spots. It is a northern species which has been taken on various conifers in both the northeastern and northwestern states and adjacent Canadian provinces.

T. latipennis V. D. has a distinctive bicolored pronotum. It is a common species on conifers in the mountains of the Pacific Northwest but presumably occurs eastward across the Canadian zone. Three additional species are known, all northern in distribution.

7a Sides of scent gland channel not or barely elevated toward end; corium not or very indistinctly punctate; lateral margins of pronotum not or very slightly flattened anteriorly but distinctly carinate .... *Anthocoris*

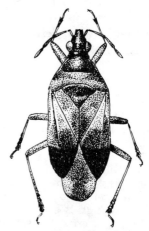

Figure 274

Figure 274 *Anthocoris musculus* (Say)

This is an elongated, relatively rather large species (3.5-3.7 mm) with the hemelytra completely shining. The anterior half of the corium is pale testaceous. The head, pronotum, inner area of the clavus and posterior half of the corium are black or very dark brown. The membrane is translucent with a dark irregular posterior area. It is most frequently found on willows and related trees. The distribution is apparently nearly country-wide across the northern states and southern Canada.

A. antevolens White is a very closely related species that is widely distributed in the western states. There is some question as to whether it actually represents a distinct species, but *antevolens* is distinguished chiefly by longer and more distinct pubescence on the hemelytra which may be expressed as having the hairs of the cuneus distinctly longer than the distance between their bases, which is not true of *musculus*. Ten to twelve additional species occur, all but one found in western North America.

REFERENCE
Carayon, J. 1972. Caracteres systematiques et classification des anthocoridae (Hemipt.). Ann. Soc. Ent. Fr. (N. S.) 8:(2):309-349.

MICROPHYSIDAE

This is a very small family of minute insects. The body is oval or oblong and flattened. The head projects straight forward. Ocelli are present. The wings are very broad with the lateral margins greatly expanded and a distinct cuneus is present. The tarsi are two-segmented. In one species the beak is three-segmented. The wing membrane has a small cell near the base and otherwise lacks venation. Only a single species occurs north of Mexico, although other rare forms considered to be somewhat intermediate between this family and the Anthocoridae are known from the Southwest.

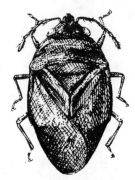

Figure 275

Figure 275 *Mallochiola gagates* (M. & M.)

This species appears to be known only from females which are glossy black and shining with a slightly paler membrane and with the hind tibiae curved and considerably longer than the femora. It is a very minute insect being only slightly over 1 mm in length. *M. gagates* is known in this country only from Maryland and the District of Columbia but it has also been reported from Mexico, so the distribution must be much more extensive.

MIRIDAE
The Plant Bugs

This is the largest family of the Heteroptera and is represented in North America by many hundreds of species. The majority of the species are delicate, fragile insects that are sometimes brightly marked but often cryptically colored in greens and brown. Most species occur on vegetation, the majority feeding upon plants but many are predaceous upon other small insects. Mirids may be recognized by the presence of a cuneus in the wing and one or two cells rather than elongate veins located near the base of the membrane of the front wing (fig. 276). Ocelli are never present. The antennae are four-segmented and frequently the third and fourth segments are more slender than the first two and thread-like in appearance.

Figure 276

The identification of specimens in such a large family as this is a real challenge to the beginning student. A great deal of additional classificatory work remains to be done in this country, and the biology and ecology of most of the species is very poorly understood. Many of our species have a single generation a year and can only be collected in the late spring and early summer when vegetation is succulent and actively growing. Any student who carefully studies this family in a given locality is certain to add many new facts of importance to science.

In the following key we have departed from our practice in the remainder of the book in that we have attempted to include all of the North American mirid genera, but to severely limit the number of species discussed.

It has seemed to us important to enable the serious student to identify Miridae to genus. At the same time we recognize that it is beyond the scope of this work to enable any significant portion of the species to be identified.

The key below should be approached with some understanding of its limitations. (1) Specimens must be in reasonably good condition since many couplets in the key require the use of the types of hairs that are present on the body surface and these are easily rubbed off. (2) It is necessary to have available a good binocular dissecting microscope with quality illumination as the major groups of Miridae are separated from one another by the condition of the arolia (=parempodia) and pseudarolia (=pulvilli) situated between the tarsal claws, and these are very difficult to see without good optical equipment. (3) A number of mirid genera are recently described and representatives have not been available to us for study. We have also not been able to examine representatives of a few of the older genera. Such genera are included in the key as accurately as we have been able to place them from a study of the literature. (4) In the larger genera we have often been able to examine only a relatively small proportion of the described species. This has also been true of most of the more extensive previous keys in the literature.

It is likely that some species will not key to the correct genus, and this possibility should constantly be kept in mind. (5) Some characters used in the key are difficult to interpret without comparative material and students should attempt to compare determinations with previously authentically determined material wherever possible.

The works of Knight and Carvalho cited below are of fundamental importance in identification of Miridae and since they are profusely illustrated are valuable as aids in the use of the present key.

The distribution limits of many Miridae are very poorly known and students should expect to frequently find species and genera outside of the ranges indicated here.

The student who wishes to accurately identify his specimens to species often must be willing to dissect and study the genitalia for frequently a whole series of species are present in a genus that look very much alike but which differ remarkably in the structure of the genitalia. By attempting to enable the student to identify specimens to genus we hope to stimulate interest in this great and complex family and to offer him or her an introduction to the more technical literature necessary for accurate species determination.

Figure 277

Figure 278

Figure 279

Figure 279 *Nevadocoris bullatus* Kngt.

Figure 280

Figure 280 *Labops hirtus* Kngt.

1   Arolia present, arising between the claws. Fig. 277 2

1a Arolia absent, with only a pair of straight hairs present. Figs. 278, 279 131

2   Arolia divergent, usually dilated (fig. 277); pronotal collar frequently present and separated from rest of pronotum by a definite furrow 3

2a Arolia usually convergent (fig. 280), sometimes parallel; pronotal collar when present not separated from remainder of pronotum by a definite furrow 77

3   Pronotum lacking a distinct pronotal collar with a complete definite furrow separating collar from remainder of pronotum (posterior furrow frequently present laterally, but never complete across midline); (in *Collaria* [fig. 282] the pronotal calli are swollen to give impression of a complete collar) 4

Figure 281

Figure 281 *Calocoris* sp.

3a Pronotum with a complete anterior collar delimited posteriorly by an impressed line that is complete across meson (fig. 281), or if collar indistinct then species are ant mimics 15

4   Second antennal segment strongly clavate .... *Actitocoris*

(One rare Holarctic species, *signatus* Reuter, is known from western Canada.)

4a Second antennal segment linear or at most very slightly enlarged near distal end 5

5   Eyes far removed from anterior margin of pronotum, at least as far removed from pronotal margin as dorsal width of an eye .... *Collaria*

Figure 282

Figure 282 *Collaria meilleurii* Provancher

This is a medium-sized plant bug (6-7 mm), of a black color with yellowish front wings and yellow-brown legs. The eyes are strongly protrudent. The pronotum has a pair of dull black spots present on the posterior lobe and there is a diffuse black area near the apex of each corium. It lives in damp meadows on various grasses. The distribution is northern from New England and eastern Canada west to Alberta but it has not been reported south of Pennsylvania and West Virginia.

C. oculata (Reuter) is closely related but has a brown rather than black pronotum. It is widely distributed in the southern states but ranges northward into New England, New York and Ohio.

One additional species (*oleosa* (Dist.)) occurs in the extreme southern and southwestern states.

5a   Eyes usually in contact or nearly in contact with anterior margin of pronotum, at least distance of eye from pronotal margin less than dorsal width of an eye 6

6   Front wings, especially clavus, deeply and coarsely punctate .... *Porpomiris*

(Two species occur, one, *curtulus* (Reut.), eastern and central United States, the other, *albescens* (V.D.), in Arizona.)

6a   Front wings smooth or roughened, at most with a very few fine punctures (*Teratocoris*) 7

7   Pronotum deeply and coarsely punctate 8

7a   Pronotum impunctate or at most with a few very fine obscure punctures 9

8   First antennal segment relatively short, not or only slightly exceeding width of head .... *Stenodema*

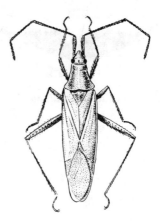

Figure 283

Figure 283   *Stenodema vicinum* (Provancher)

This is an elongate, slender straw-colored insect (often greenish when alive) with two or three dark longitudinal stripes on the pronotum. The first antennal segment is thick and densely clothed with long hairs. The antennae are a reddish color. *S. vicinum* is a very common species over much of the United States where it occurs in meadows and along roadsides on various kinds of grasses.

Figure 284

S. trispinosum Reuter (fig. 284) is very similar in appearance and habits to *vicinum* but is easily distinguished by the presence of three spines on the underside of the hind femora near the distal end. It is distributed across the entire continent in the northern states.

There are definite ecological differences in the habitats of these two species but they have not been carefully studied even in a single locality. (Four or five additional species are reported from North America but the status of some of them is in need of additional study.)

8a First antennal segment relatively elongate, exceeding width of head plus width of an eye .... *Litomiris* (Fig. 285)

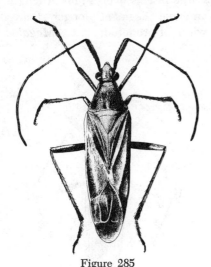

Figure 285

Figure 285 *Litomiris debilis* (Uhler)

(Seven nominal species are recognized, all western in distribution. Iowa is the most eastern state represented.)

9 First antennal segment bearing long hairs or suberect bristles, these at least 1/2 as long as width of segment. Fig. 286 10

9a First antennal segment with only very short hairs present, their length less than 1/2 width of segment 12

10 Body almost hairless; eyes nearly touching anterior margin of pronotum; vertex with a distinct longitudinal groove .... *Dolichomiris*

(A single cosmopolitan species, *linearis* Reuter, occurs in Florida.)

10a Body covered with conspicuous semi-upright hairs; eyes a little removed from anterior margin of pronotum; vertex of head lacking a definite longitudinal groove 11

11 First antennal segment clothed with erect, long pilose hairs; pronotal calli with a deep (foveate) depression on middle of each callus ...... *Chaetofoveolocoris*

(A single species, *hirsuta* (Kngt.), is known from Arizona and Texas.)

11a First antennal segment clothed with suberect stiff bristles; calli lacking a deep foveate pit but often with an irregular depressed area present .... *Leptopterna*

Figure 286

Figure 286 *Leptopterna dolabrata* (L.)

This is the common "Meadow Plant Bug" that swarms in vast numbers in meadows and pastures in spring and early summer. It sometimes does considerable damage to blue grass. It is a large species, 7-9 mm in length. The color is variable. Males are often orangish yellow to red-brown, other males and females are pale greenish white with two black

Families of Hemiptera   151

stripes on the pronotum. Females are often short-winged. *Dolabrata* occurs widely over the northern United States into which it may have been introduced many years ago from Europe.

One additional species, *ferrugata* (Fallen), has been reported from this country.

12   Head short and blunt with apex evenly rounded and scarcely extending forward beyond bases of antennae .... *Teratocoris*

Figure 287

Figure 287   *Teratocoris discolor* Uhler

This species has a pale greenish ground color with the head, the central area of the pronotum and a large central area on the front wings dark brown to black. The legs and first two antennal segments are reddish. It is 4.5-5 mm long. *T. discolor* occurs in the northern and central states and Canada from Quebec to British Columbia and south to Missouri and New York. Five additional northern species have been reported from North America.

12a Head pointed and rather elongate, extending a considerable distance beyond bases of antennae 13

13   Pronotum with a well-defined median, raised, carinate, longitudinal ridge; head lacking a deep longitudinal groove on vertex but with a depressed area present .... *Acetropis*

(One species, *americana* Kngt., is known from Oregon.)

13a Pronotum lacking a distinct carinate median ridge (an inconspicuous median elevation often present), a distinct longitudinal groove present on vertex 14

14   Large insects over 8 mm in length; first antennal segment longer than basal width of pronotum .... *Megaloceroea*

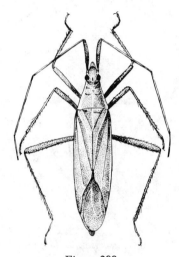

Figure 288

Figure 288   *Megaloceroea recticornis* (Geoff.)

This is a large elongate, nearly uniformly green plant bug that has been introduced from Europe. It occurs in large numbers on quack grass and related species. The distribution, as with many other recently introduced species, is spotty but it is widespread in the northeastern and north central states.

14a Smaller insects, usually less than 7 mm in length; first antennal segment shorter than width across base of pronotum .... *Trigonotylus*

152   Families of Hemiptera

Figure 289

Figure 289 *Trigonotylus coelestialium* Kirkaldy

This is a small (5-6 mm) grass-green insect with red antennae. The body is elongate and slender. The pronotum is obscurely marked with four dusky longitudinal stripes. It feeds upon various wild grasses and oats in pastures, meadows and roadsides. It is most abundant in the northern states and Canada.

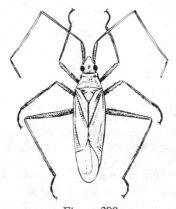

Figure 290

T. tarsalis (Reuter) (fig. 290) is similarly colored but has distinct black tarsi and tibial apices. It is widely distributed on *Spartina* grasses in marshy areas.

Approximately fifteen additional North American species are known, several of them widely distributed.

15 Basal segment of posterior tarsus as long as segments two and three combined; body form somewhat ant-mimetic with membrane undifferentiated or vestigal. Fig. 291 16

15a Basal segment of posterior tarsus shorter than segments two and three combined, if subequal then not ant-mimetic and with cuneus and membrane well developed 17

16 Length of first antennal segment less than width of vertex. Fig. 291 .... *Pithanus*

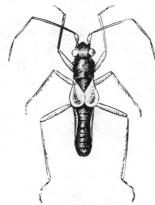

Figure 291

(One species, *maerkelii* (H. S.) (fig. 291), occurs across the extreme northern states and southern Canada from Nova Scotia to British Columbia.)

16a Length of first antennal segment greater than width of vertex. Fig. 292 .... *Mimoceps*

Figure 292

(One mimetic species, *insignis* Uhler (fig. 292) occurs on sedges and marsh grasses across the northern states and Canada.)

17 Ant-mimetic; abdomen strongly constricted laterally at base (fig. 293); pronotal collar represented only by a depressed line 18

Figure 293

Figure 293 *Barberiella apicalis* Kngt.

17a Species not ant-mimetic, abdomen not strongly constricted at base; anterior pronotal collar well developed 21

18 Scutellum bearing an erect spinelike projection .... *Barberiella*

(Two rare species occur in North America.)

18a Scutellum lacking an upright spinelike projection 19

19 Pronotal width across humeri greater than head width across eyes; hind femora with numerous upstanding elongate hairs; second antennal segment completely linear .... *Paraxenetus*

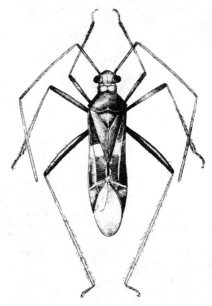

Figure 294

Figure 294 *Paraxenetus guttulatus* (Uhler)

This is a large (6.5-7 mm) grayish-brown mirid often marked with red on the cuneus, membrane veins and spots on the legs. It occurs chiefly on grape vines in the southern and central states north to New York and west to Iowa and Texas.

19a Width of head across eyes conspicuously greater than width of pronotum across humeri; hind femora lacking elongate upstanding hairs; second antennal segment somewhat swollen distally 20

20 Pronotum raised posteriorly into an erect spinelike projection. Fig. 295 ....
.... *Dacerla*

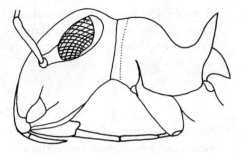

Figure 295

Figure 295 *Dagerla mediospinosa* Sign.

(A strongly ant-mimetic species occurs in California.)

20a Pronotum posteriorly lacking a raised spinelike projection. Fig. 296 .... *Paradacerla*

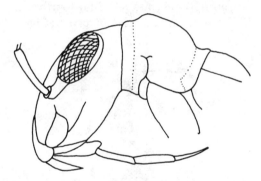

Figure 296

Figure 296 *Paradacerla formicina* (Parsh.)

(Two ant-mimetic species occur in the far western states.)

21 Scent gland orifice and evaporative area small and indistinct, dorsal margin not extending above ventral margin of mesepimeron. Fig. 297 22

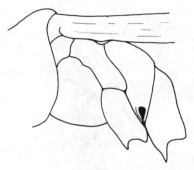

Figure 297

Figure 297 *Opistheurista clandestina* (V.D.)

21a Scent gland orifice well developed, its dorsal margin extending well above ventral margin of mesepimeron. Fig. 298 .. 23

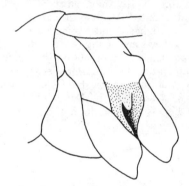

Figure 298

Figure 298 *Lygus lineolaris* (P.B.)

22 First antennal segment shorter than width of vertex; second antennal segment more than three times length of first .... *Oncerometopus*

(Eight species occur, chiefly in the western United States.)

22a First antennal segment longer than or subequal to width of vertex; second antennal segment frequently not more than twice length of first (but sometimes more than three times length of segment one). Fig. 299 .... *Prepops*

Families of Hemiptera 155

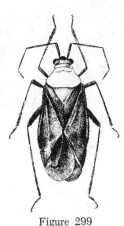

Figure 299

Figure 299 *Prepops insitivus* (Say)

Twenty-three North American species occur, many widely distributed. Most species are of medium or large size with heads and pronota variously marked in reds, yellows and oranges and with dark wings and appendages. The pronotum is strongly sloped downward from the posterior lobe. One common species (*fraternus* Kngt.) breeds on sumac. *Opistheurista clandestina* (V.D.) (fig. 300) is included here since we do not find reliable distinguishing characteristics.

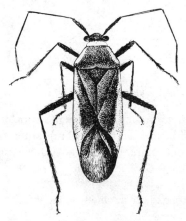

Figure 300

Figure 300 *Opistheurista clandestina* (V.D.)

23   Pronotum strongly punctate or if punctures appear faint then surface rugulose or has scalelike hairs or a silky hair covering present 24

23a   Pronotum lacking distinct punctures or if faint punctures present then surface is neither rugulose nor in this case does it bear either scalelike hairs or silky hairs. 52

24   Second antennal segment strongly thickened toward distal end (clavate) .. 25

24a   Second antennal segment linear or nearly so, never strongly swollen at distal end 28

25   Length of second antennal segment greater than twice width of head; antenal segment two slender on proximal half, abruptly clavate on distal fourth .... *Ectopiocerus* (Fig. 301)

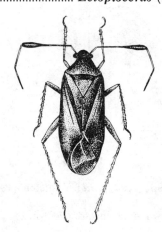

Figure 301

(One species, *anthracinus* Uhler (fig. 301) occurs in California.)

25a   Length of second antennal segment less than twice width of head; antennal segment two relatively gradually thickened from proximal half to distal end 26

26 Juga prominently and abruptly convex; body frequently entirely black (color varieties with a pale head and pronotum occur) .... *Capsus*

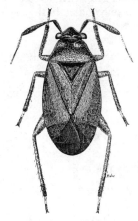

Figure 302

Figure 302 *Capsus ater* (L.)

This is generally a uniformly black mirid with a broad elliptical shape which is easily recognized by the greatly swollen distal portion of the second antennal segment. It is often abundant on blue grass and other grasses along roadsides. Color varieties occur with yellow legs and also with a reddish or orange head and pronotum. These color morphs have been given Latin names but appear to have no geographical significance and use of such names should be avoided. This is apparently an introduced European species and is widely distributed.

(One additional species, *simulans* (Stal), occurs in North America.)

26a Juga not prominently and abruptly convex 27

27 Second antennal segment equal to or longer than width across base of pronotum .... *Notholopus*

(One species, *californicus* Kngt., occurs in California.)

27a Second antennal segment shorter than width across base of pronotum .... *Pycnocoris*

(One species, *ursinus* V.D., occurs in California.)

Figure 303

Figure 303 *Derophthalma* sp.

28 Scutellum strongly swollen and produced (fig. 303); front wings covered thickly with small scalelike hairs .... *Derophthalma*

(One species, *variegata* (Blatchley), is found in Florida.)

28a Scutellum usually not strongly swollen and produced (in the few cases where the scutellum is somewhat swollen front wings are not covered with scalelike hairs) 29

29 Front wings clothed with distinct woolly hairs lying flat against body and intermixed with fine upright hairs .... *Calyptodera*

(One species, *robusta* V.D., occurs in Lower California.)

29a Front wings either hairless or with only a single kind of hair present, these, when present, usually not woolly (frequently woolly in *Irbisia*) 30

Figure 304

Figure 304 *Tropidosteptes glaber* (Kngt.)

30 Pronotum distinctly punctate between and in front of calli. Fig. 304 31

30a Pronotum not distinctly punctate before and between calli (sometimes roughened in this area) 33

31 Form rather ovoid, hemelytra not parallel-sided; lateral pronotal margins frequently carinate or at least calloused .... *Tropidosteptes*[3]

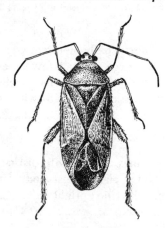

Figure 305

Figure 305 *Tropidosteptes cardinalis* Uhler

This is a bright red insect with the antennae, tip of head, wing membrane and legs strongly contrastingly black. It is about 5.5 mm in length and lives primarily on white ash from New England to Florida east of the Great Plains.

  T. amoenus (Reuter) is another common species on ash trees east of the Great Plains. It is a smaller species (4.25-5 mm) of a pale yellow color marked with a series of brownish stripes on the pronotum. The apex of the cuneus is usually bright red.

  This is a rather large genus containing twenty-five additional North American species, and is represented throughout North America chiefly where ash grows.

31a Body shape more elongate and nearly parallel; lateral margins of pronotum not carinate 32

32 Second antennal segment four (females) or five (males) times as long as segment one; body not over 4 mm in length .... *Neoborella*

(Two species occur in Colorado and Arizona.)

32a Second antennal segment not more than three times as long as segment one; body over 4 mm in length .... *Xenoborus*

(Six widely distributed and often locally common species occur most commonly on ash east of the Great Plains.)

33 Body hairless and shining, scutellum smooth; labium extending backward at least to distal end of hind coxae 34

33a Body clothed with hairs (if a very few hairs are present then scutellum is either roughened or punctate); labial length variable 36

3. The generic limits are unsatisfactory, some species currently placed in *Xenoborus* will run to here and vice versa. A revisional study of the complex is needed.

34   Labium elongate, reaching onto fourth or fifth segments of abdomen, front wings translucent; vertex with cross ridges or markings (striolate). Fig. 306 ....
.... *Platylygus*

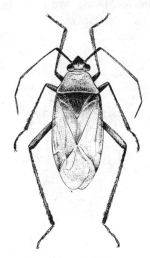

Figure 306

Figure 306   *Platylygus grandis* Kngt.

(A genus of six large yellowish-brown species mostly southwestern in distribution with one rare species known from the northeast.)

34a   Labium extending at most only to distal ends of hind coxae; front wings at most only very weakly translucent; vertex without cross ridges or markings 35

35   Labium reaching distal ends of hind coxae .... *Proba*

(Three species occur in the southwestern states.)

35a   Labium reaching only to middle coxae .... *Rhasis*

(One species, *laeviscutatus* (Kngt.) occurs from North Carolina and Illinois west to Texas and Oklahoma.)

36   First and second antennal segments thickly clothed with prominent black or dark brown hairs; body color chiefly bright red .... *Coccobaphes*

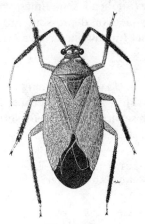

Figure 307

Figure 307   *Coccobaphes sanguinareus* Uhler

This is a very handsome bright red plant bug with the wing membrane and the first and second antennal segments a strongly contrasting black. This species occurs on sugar and red maples throughout the northeastern and central United States and has been taken in Florida.

36a   First antennal segment having at most a sparse clothing of hairs present, these usually pale, if black hairs present then body chiefly or entirely black 37

37   Lateral margins of pronotum sharply angulate with a distinct ridge near humeral angles; clavus without punctures; color red or orange and black .... *Neocapsus*

(A single species, *cuneatus* Dist., is found in the southern and southwestern states.)

37a   Lateral pronotal margins either rounded or angulate, if somewhat carinate then clavus has at least two rows of punctures; color variable but rarely red and black 38

Families of Hemiptera   159

38   Vertex of head with a distinct median longitudinal groove .... *Creontiades* (See couplet 67)

38a Vertex of head sometimes with an obscure median depression but lacking a distinct furrow-like groove 39

39   Body black, usually shining or subshining and with woolly pubescence; jugum and lorum separated by a deep suture .... *Irbisia*

(Fifteen species of these large black insects occur in the western states, several frequently very common.)

39a Body variously colored, frequently mottled brown and yellow but also red or green, rarely completely black but if so then lacking a deep suture between jugum and lorum 40

40   Pronotum distinctly and coarsely roughened (rugose) with punctures usually obscured by roughened surface 41

40a Pronotum at most very finely roughened, never so strongly so as to obscure punctures 42

41   Pronotal calli large, coming together at midline and reaching pronotal margins; vertex width greater than width of an eye .... *Plesiocoris*

(One species, *rugicollis* (Fallen), has been taken in Alberta, British Columbia and Alaska.)

41a Pronotal calli small, not in contact at midline and not attaining lateral margins of pronotum; width of vertex less than width of an eye .... *Neoborops*

(A single species, *vigilax* Uhler, is known from Colorado and Arizona.)

42   Second antennal segment only as long as, or shorter than, width of head across eyes .... *Agnocoris*

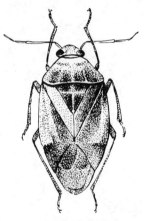

Figure 308

Figure 308   *Agnocoris pulverulentus* (Uhler)

This is a small dark yellowish-brown mirid (4.5 mm) that occurs on willows. It is distributed over most of the United States but its ecological requirements and relationships to the various species of willows are poorly understood. Three additional species occur in North America.

42a Second antennal segment longer than width of head across eyes 43

43   Calli naked, lacking fine hairs 44

43a Calli clothed with hairs (orient laterally— sometimes very sparsely so) 45

44   Lateral margins of front wings nearly parallel sided, shape elongate; cuneus only moderately deflexed .... *Lygidea*

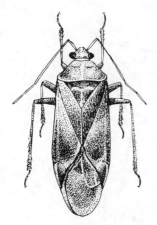

Figure 309

Figure 309 *Lygidea mendax* Reuter

This is an orange-red insect with the clavus, basal half of the corium and often the base of the pronotum dark brown to black. This mirid is known as the "Apple Red Bug" and was once a serious pest of apples in the northeast, but its injury has been reduced to minor proportions in recent years. It also feeds on hawthorn, crab apple and quince. The range is through eastern Canada, the northeastern states and west to Iowa. Eight additional species are found widely distributed throughout at least the northern United States and southern Canada.

44a Lateral margins of front wings moderately convex, not parallel sided; cuneus strongly deflexed .... *Lygus*[4]

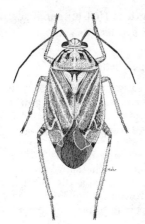

Figure 310

Figure 310 *Lygus lineolaris* (P. B.)

This is the ubiquitous tarnished plant bug and is one of the most abundant of all Hemiptera in the eastern and central states. It is a yellowish-brown insect with variable amounts of brown or black mottling on the body. The scutellum usually has a conspicuous yellow Y-shaped marking and the pronotum has several more or less complete longitudinal dark rays. It ranges over most of the eastern and central states and is often very destructive to legumes, fruit, vegetables, and flowers. The tarnished plant bug has a number of wild hosts but seems to prefer plants belonging to the families Compositae and Umbelliferae.

There are a great many (approximately 46) very closely related species in this genus and their accurate identification depends upon the use of characters found in the male and female genitalia. A critical study of these populations using modern population concepts would be of great value as some "species" look suspiciously like geographic races. Some western "lygus-bugs" are serious crop pests.

45   Pronotum deeply and thickly punctate, hairs nearly flat against body surface .. 46

45a Pronotum relatively shallowly, finely and sparsely punctate; hairs long, fine and suberect 48

46   Collar of pronotum wider than width of second antennal segment .... *Lygus* (See couplet 44a)

46a Collar of pronotum narrower than or only as wide as width of second antennal segment 47

47   Frons of head smooth; (size relatively small, not exceeding 5.5 mm) .... *Orthops*

Orthops scutellatus (Uhler)

This is a small species (4 mm) of a greenish yellow color marked with brown. The scutellum is a

4. *Knightomiris* Kelton probably will key near this couplet.

Families of Hemiptera

bright green or yellow. It occurs across the northern states and southern Canada on poison hemlock, wild parsnip and other umbelliferous plants and has been reported injuring celery.

47a Frons of head grooved or ridged (striolate) .... *Salignus*

Figure 311

(A single western species, *distinguendus* (Reuter) [fig. 311] is known.)

48 Labium extending posteriorly to fifth or sixth abdominal segment 49

48a Labium not extending posteriorly beyond ends of hind coxae 50

49 Second antennal segment very elongate, twice or nearly twice as long as width of head across eyes; fore femora not conspicuously banded or spotted with red; pronotal punctures on posterior lobe unicolorous with remainder of surface .... *Pinalitus*

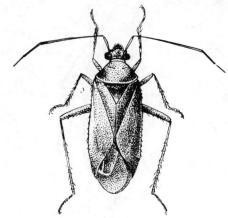

Figure 312

(One species, *approximatus* (Stål) (fig. 312) occurs on goldenrod across the northern states and in southern Canada. Five additional southwestern species are known.)

49a Second antennal segment not more than 1 1/4 longer than width of head across eyes; fore femora distally with conspicuous dark red spots or bands; punctures on posterior pronotal lobe dark brown contrasting with pale surrounding area .... *Camptozygum*

(One introduced species, *aequale* (Vill.), has recently been reported from Pennsylvania.)

50 Length of first antennal segment greater than eye length .... *Lygocoris*

Lygocoris pabulinus (L.)

This is a nearly uniformly pale green mirid (5.5-6 mm) that occurs over the northern states and southern Canada. It is often common on Touch-me-not (*Impatiens biflora*) in wet places along roadsides.

A very large number (36) of northeastern species belong to the subgenus *Neolygus* (sometimes considered as a distinct genus). Most of these are brownish or greenish insects and are usually host specific to many of our native trees and shrubs. Identification is often very critical, involving the use of genitalic characters. Among the more common species are *communis* Kngt. (fig. 313) on dogwoods and pear, *quercalbae* Kngt. on oak, *tiliae* Kngt. on linden, *caryae* on hickory and *omnivagus* Kngt. on oaks. This subgenus would well repay careful ecological analysis to ascertain specific limits and the effect of environmental variations on host selection and morphological appearance. Two members of the subgenus *Apolygus* (sometimes considered to be a separate genus) are also found north of Mexico.

(Two species are found in the eastern and central United States.)

Figure 314

52 First antennal segment thickly clothed with numerous flattened hairs. Fig. 314 .... *Neurocolpus*

Neurocolpus nubilus (Say)

This is a rather large plant bug (6.5-7 mm) of a bright reddish tan to yellow-brown coloration. The second antennal segment is slightly thickened at the distal end. The front wings have many minute yellow dots scattered over the surface. The peculiar flattened hairs on the first antennal segment are diagnostic. *N. nubilus* is a common species over most of eastern and central North America. The buttonbush is one of the host plants, but it may occur on other plants as well. Thirteen additional species (including *tiliae* Kngt. on basswood) (fig. 315) occur in our territory.

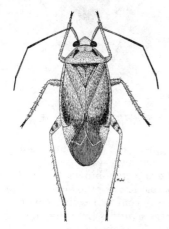

Figure 313

Figure 313 *Lygocoris communis* (Kngt.)

50a Length of first antennal segment equal to or less than eye length 51

51 Third antennal segment longer than head length .... *Taylorilygus*

Taylorilygus pallidulus Blanchard

This is a nearly uniformly green species resembling *Lygocoris pabulinus* (L.) but smaller (4.5-5 mm). Some specimens are marked with red and brown. It breeds on fleabane and almost certainly on many other plants as well. It is widespread over much of the United States but much more common in the south and is probably the most abundant plant bug in southern Florida.

51a Third antennal segment shorter than head length .... *Dagbertus*

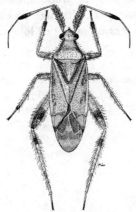

Figure 315

Figure 315 *Neurocolpus tiliae* Kngt.

Families of Hemiptera   163

52a First antennal segment lacking distinctly flattened hairs, but often with prominent hairs present 53

53 First antennal segment strongly flattened. Fig. 316 .... *Lampethusa*

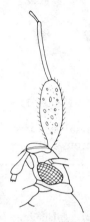

Figure 316

(Two scarce species occur in the southwest.)

53a First antennal segment rounded 54

54 Pronotum marked on anterior half by a pair of conspicuous black spots 55

54a Anterior lobe of pronotum lacking a pair of black spots, although black spots may be present on posterior lobe 57

55 First antennal segment bearing long black hairs, these as long as or longer than diameter of segment .... *Taedia*

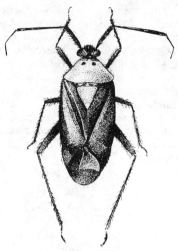

Figure 317

Figure 317 *Taedia scrupeus* (Say)

This is a rather large plant bug, usually orange and black but extremely variable in color, sometimes yellow or completely brown. The tibiae bear very long hairs that are as prominent as the true spines and are easily confused with them. This character will separate *scrupeus* from the other members of the genus. The species lives on wild grapes and probably other plants as well. Many color varieties have been given formal names but so far as known these do not have geographic significance. This species ranges over the United States west to Colorado and Texas.

Over twenty-five additional species have been described from North America. The genus is badly in need of revisional study.

55a First antennal segment with only a few scattered hairs and these never as long as thickness of segment 56

56 Dorsal surface bearing distinctly silky or woolly pubescence .... *Polymerus* (See couplet 76a)

56a Dorsal surface bearing only simple hairs. Fig. 318 .... *Calocoris* (See couplet 75)

Families of Hemiptera

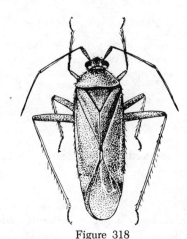

Figure 318

Figure 318 *Calocoris norvegicus* (Gmelin)

57 Second antennal segment thickened and rather spindle shaped .... *Garganus*

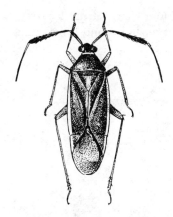

Figure 319

Figure 319 *Garganus fusiformis* (Say)

This is a very handsome small mirid (5 mm) which is chiefly black in color marked with white on the anterior collar of the pronotum, as a broad median stripe through the center of the scutellum, along the claval commissure and on the outer margins of the front wings. The first antennal segment and the legs are orange, the second antennal segment is spindle shaped and black, the third segment black with a yellow area at the base. It breeds on Coltsfoot (*Tussilago farfara*). The distribution is over the entire eastern states west to Kansas.

Two additional rare species occur in the southwest.

57a Second antennal segment linear, or at most very slightly enlarged distally, but never distinctly enlarged and spindle shaped 58

58 Hind femora long, extending considerably beyond end of abdomen, flattened, tapering (if femora only slightly exceeding abdomen then first segment of hind tarsus not longer than third, and length of first antennal segment greater than length of head and pronotum combined) 59

58a Hind femora not or only very slightly exceeding end of abdomen, not flattened or strongly tapering 60

59 Color creamy white, lateral pronotal margins with a distinct acute ridge (carinate); a median line of white hairs running through head, pronotum and scutellum .... *Pallacocoris*

(One uncommon species, *suavis* Reut., is known from Iowa and Texas.)

59a Color variable, usually brown and black with yellow or white spots present; never having a median line of white hairs and lateral pronotal margins not sharply ridged. Fig. 320 .... *Phytocoris*

Families of Hemiptera

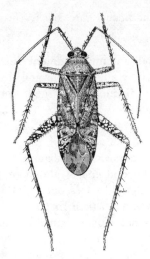

Figure 320

Figure 320 *Phytocoris salicis* Kngt.

This is one of the largest and most complex of all the North American mirid genera. One hundred and ninety-eight species are known from our area. Many of them are bewilderingly similar in general appearance and their determination requires careful dissection of the genitalia and sometimes authenticated specimens for comparison. Many species are reasonably common and some come to light.

60   First segment of hind tarsus distinctly longer or subequal in length to third .... *Stenotus*

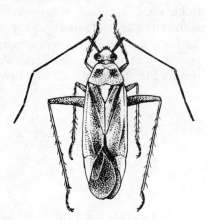

Figure 321

Figure 321 *Stenotus binotatus* (F.)

This is a very common yellowish orange slender species (6 mm) with two broad black stripes (often reduced to spots on posterior lobe) on the pronotum and on the front wings. It occurs throughout the northern and central United States on various grasses and is often very common in meadows and along roadsides.

60a First segment of hind tarsus shorter than third 61

61   Dorsal body surface without hairs or nearly so 62

61a Dorsal surface distinctly covered with at least short hairs 64

62   Labium short, at most barely extending posteriorly beyond front coxae .... *Poecilocapsus*

Figure 322

Figure 322 *Poecilocapsus lineatus* (F.)

This is a common brightly colored large ovoid mirid (7-7.5 mm) known as the "Four-lined Leaf Bug." The ground color is green in life, fading to yellow, with four broad black stripes or bands on the pronotum and wing covers. The head is often yellowish red. It breeds upon a large number of plants, sometimes becoming injurious in gardens, especially to currants, daisies, dahlia, etc. It ranges throughout the eastern states west to the Great Plains and in the north to British Columbia. A second species has been reported from Florida and California.

62a Labium longer, extending posteriorly at least to posterior margin of middle coxae 63

63 First antennal segment as long as length of pronotum .... *Ganocapsus*

(One species, *filiformis* V.D., is known from Arizona.)

63a First antennal segment much shorter than length of pronotum .... *Metriorhynchomiris*

Metriorhynchomiris dislocatus (Say)

This is a rather large ovoid robust mirid (6.5 mm) of a shining aspect and extremely variable in color with fifteen different color varieties described but none as yet having been shown to have geographic significance. The most common coloration is reddish orange with large spots on the basal area of the scutellum, a stripe on the corium, and the wing membrane black. It breeds on False Solomon's Seal, Wild Geranium and sometimes other plants, usually in moist shaded places. The distribution is from the northeastern states west to Minnesota and south to Texas in the west.

M. fallax Reuter is a somewhat smaller less ovoid species black in color with a reddish or yellowish scutellum (sometimes also black) and yellow legs. It is found east to Pennsylvania and west to Iowa on Redbud and wild Gooseberry. One additional species is known in the United States.

64 Head subvertical, vertex somewhat flattened; width of head across eyes almost as great as width of pronotum across base. Fig. 323 .... *Bolteria*

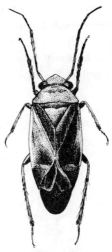

Figure 323

Figure 323 *Bolteria* sp.

(Fourteen species occur, chiefly in the western and southern states.)

64a Head declivent, vertex convex; width of head across eyes distinctly less than width of pronotum across base 65

65 Pronotum strongly carinate to explanate laterally .... *Allorhinocoris*

(Two species occur in the western states.)

65a Pronotum laterally sometimes subcarinate but never sharply so and never appearing explanate 66

66 Hemelytra with black setiform hairs and silky silvery pubescence present; labium very elongate reaching eighth abdominal segment .... *Ecertobia*

(One species, *decora* Reuter, occurs in the western states.)

Families of Hemiptera   167

66a Hemelytra with only a single type of pubescence present which may either be simple, woolly or black and setiform; labium not nearly attaining eighth abdominal segment 67

67 First antennal segment longer than width of head; vertex with a distinct median longitudinal sulcus. Fig. 324 .... *Creontiades*

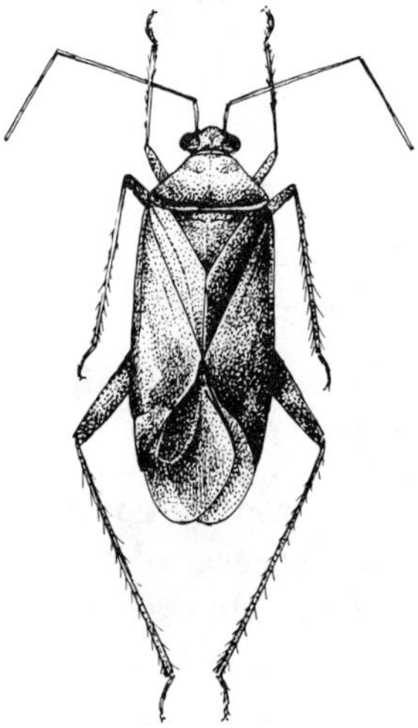

Figure 324

Figure 324 *Creontiades* sp.

(Three species occur in southern and western North America.)

67a First antennal segment subequal to or shorter than head width; vertex usually lacking a distinct longitudinal sulcus 68

68 Females brachypterous; males with cuneus two and one-half times as long as wide; tibiae with very elongate spines .... *Stittocapsus*

(One species, *franseriae* Kngt., occurs in the southwestern states.)

68a Females usually macropterous, if brachypterous females occur then males with cuneus less than two and one-half times as long as wide; tibiae generally with normally developed non-elongate spines 69

69 Color black with jugum and lorum separated by a deep suture; antennal socket located below ventral margin of eye. Fig. 325 .... *Irbisia*
(See couplet 39)

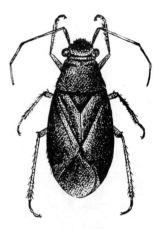

Figure 325

Figure 325 *Irbisia* sp.

69a Color usually other than black, if black then jugum and lorum not separated by a deep suture and antennal socket located well above ventral margin of eye 70

70 Body above bearing stout black setiform bristles .... *Miris*

(One species has been reported, somewhat questionably, from North America.)

70a Body above lacking stout setiform bristles 71

71 Posterior margins of eyes somewhat flattened, forming an even arcuate line with base of head; labium generally surpassing distal ends of posterior coxae 72

71a Eyes convex on posterior margins, not forming an even arcuate line with base of head; labium not surpassing posterior coxae 73

72 Width of head across eyes distinctly less than width of pronotum across base; vertex of head convex .... *Dichrooscytus*

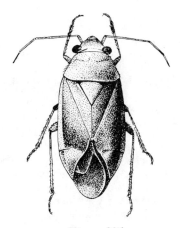

Figure 327

D. elegans Heid. (fig. 327) is a closely related species which often occurs with *repletus* but differs in that it has reddish front wings. Twenty-eight additional species occur in North America.

72a Width of head across eyes nearly as great as width of pronotum across base; vertex of head flat .... *Bolteria*
(See couplet 64)

73 Mesal length of pronotal collar nearly equal to diameter of fourth antennal segment .... *Adelphocoris*

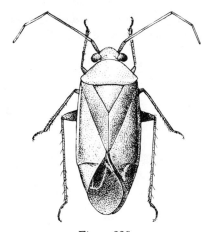

Figure 326

Figure 326 *Dichrooscytus repletus* (Heid.)

This is a small uniformly rich green insect (3.25 mm) with the third and fourth antennal segments dark. It is often common on Red Cedar and Arbor Vitae. It occurs widely in the eastern states. Most literature is under the name *viridicans* Kngt.

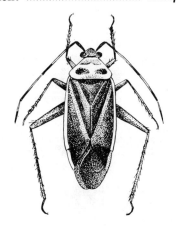

Figure 328

Figure 328 *Adelphocoris rapidus* (Say)

This common species is known as the Rapid Plant Bug. It is a large mirid (7-8 mm) of a dark brown color with dull pale yellowish margins to the front wings. The pronotum has two dark spots (often fused mesally) on the posterior half. The antennae are alternately black and light yellow (segment one, distal 1/3 of two, distal 3/4 of three and four black). It breeds chiefly on dock but sometimes injures cotton and legumes. It is often a very abundant species. *A. rapidus* occurs everywhere east of the Great Plains. The status of *superbus* (Uhl.), a variable often red "species" which replaces *rapidus* in the west, needs further study to determine if the two are indeed separate species.

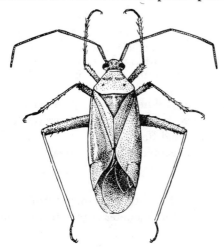

Figure 329

A. lineolatus (Geoze) (fig. 329) is a slightly larger green or greenish-yellow species with vague dusky markings on the front wings. Many of our mirids appear to have been introduced from Europe, but this is our best documented case. The chief introduction was on legumes at Ames, Iowa. It has now spread widely over the country as a pest of alfalfa and sweet clover. It is known as the "Alfalfa Plant Bug" and is often a very abundant insect.

73a Mesal length of pronotal collar distinctly greater than diameter of fourth antennal segment 74

74 Pronotum glabrous; hemelytral pubescence very short .... *Ganocapsus* (See couplet 63)

74a Pronotum pubescent; hemelytral pubescence well developed never unusually short 75

75 Dorsal surface bearing only simple hairs .... *Calocoris*

(Four or five species occur chiefly in the northern and western states.)

75a Dorsal surface bearing distinctly silky or woolly pubescence 76

76 Wing membrane marked with numerous white spots .... *Pachypeltocoris*

(One species, *conspersus* Kngt., is known from Missouri.)

76a Wing membrane lacking numerous white spots .... *Polymerus*

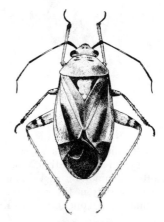

Figure 330

Figure 330 *Polymerus basalis* (Reuter)

This is a small species (3.75-4.75 mm) yellowish or tan in color with a red cuneus and reddish-brown clavus and posterior half of corium. The first antennal segment is black, with the base and distal end white. The femora are usually marked near their distal ends with a pair of dark bands.

This is a very common species over much of the country east of the Great Plains along roadsides and in fields where it is found on ragweed, dog fennel and other weeds.

Polymerus is a very large genus with forty-one additional described North American species. Many of these may be found on bedstraw, goldenrod, asters and related plants.

77   Ant mimetic species with abdomen strongly constricted at base; frequently with transverse stripes of silvery hairs 78

77a   Non-mimetic species, abdomen not strongly constricted at base; without transverse stripes of silvery hairs 87

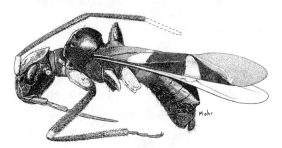

Figure 331

Figure 331   *Cyrtopeltocoris illini* Kngt.

78   Scutellum posteriorly bearing a conically produced process. Fig. 331 .... ***Cyrtopeltocoris***

(Ten rare species are present north of Mexico.)

78a   Scutellum posteriorly not having a conically produced process, sometimes mesally convex 79

Figure 332

79   Antennal segments two, three and four subequal in thickness (fig. 332) (females frequently with short or beetle-like wings) 80

79a   Second antennal segment thicker, at least on distal third, than segments three and four (fig. 333) (females usually long winged) 83

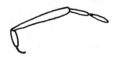

Figure 333

Figure 333   *Pilophorus* sp.

80   Front wings lacking either scaly hairs or silvery pubescent bands (although silvery markings not obviously formed of hairs are often present) .... ***Sericophanes***

80a   Front wings having flattened scalelike hairs or transverse silvery pubescent bands 81

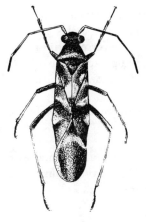

Figure 334

Families of Hemiptera   171

Figure 334 *Sericophanes heidemanni* Poppius

This is a small chestnut colored ant-like insect (3.5 mm) with two silvery bands on the front wings and a whitish spot on each clavus. It is a ground-living species found in grassy areas and frequently associated with ants. Sometimes it comes to lights in considerable numbers. It is widely distributed across the northern states and south in the east at least to North Carolina.

Seven additional closely related species occur in North America.

81   Pronotum covering mesoscutum and part of scutellum .... *Renodaeus*

(One species, *texanus* Kngt., occurs in Texas.)

81a Pronotum not covering mesoscutum and part of scutellum 82

82   Short-bodied blackish species; both sexes with wings reduced to short pads; scale-like pubescence scattered evenly over hemelytra .... *Anapus*

(One species, *americanus* Kngt., is known from Washington and Utah.)

82a Elongate, slender, varicolored species; with at least males macropterous; distinct transverse stripes of silvery pubescence present across hemelytra .... *Pilophoropsis*

(Three species occur in the southwest.)

83   Vertex of head not compressed posteriorly and not overlapping anterior margin of pronotum 84

Figure 335

83a Vertex noticeably compressed posteriorly (fig. 335), usually at least slightly overlapping anterior margin of pronotum 85

84   Scutellum strongly swollen and elevated .... *Cyphopelta*[5]

(One species, *modesta* V. D., occurs in California.)

84a Scutellum convex but not strongly elevated .... *Pseudoxenetus*

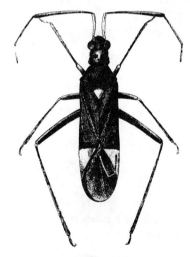

Figure 336

5. Knight (1968) keys this genus in the Mirini to which the male genitalia relate it.

Figure 336 *Pseudoxenetus scutellatus* (Uhler)

This is a large (6.5 mm) black ant-like species with a strongly contrasting yellowish-white scutellum and a white band across the base of each cuneus. It occurs on oaks from the northeastern states west to Minnesota and south to Virginia.

P. regalis (Uhler) is a closely related species that is found in the southern states. It has the posterior half of the pronotum and the lateral and ventral portions of the thorax red.

85 Front wings having white or silvery pubescent cross bands. Figs. 337 and 338 .... *Pilophorus*

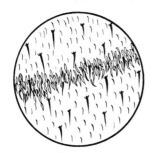

Figure 337

Figure 337 *Pilophorus* sp.

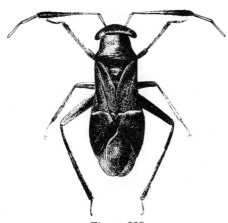

Figure 338

Figure 338 *Pilophorus* sp.

This genus contains forty-five North American species of closely related ant-like insects that are usually chocolate-brown or reddish brown in color. Most species occur on various species of pines but species also occur on cypress, larch, cedar, honey locust, willow, oak, juniper and other trees. The members of the genus are probably predaceous but much work needs to be done upon the biology of these interesting insects.

85a Front wings lacking silvery or white pubescent cross bands 86

86 Width of head equal to or wider than basal width of pronotum .... *Alepidia*

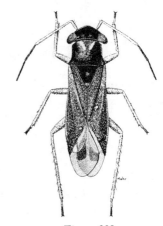

Figure 339

Figure 339 *Alepidia gracilis* (Uhler)

This is a very broad-headed black or very dark brown species (4.25 mm) with strongly contrasting yellow legs and antennae. It is sometimes common on various species of pines. The distribution is over most of the eastern and central United States west to Colorado. A second species, *bellula* Hussey, occurs in Connecticut and Michigan.

86a Width of head less than width of pronotum across base; (second antennal segment noticeably enlarged toward distal end) .... *Alepidiella*

(One species, *heidemanni* Poppius, has been reported from Maryland, Washington, D. C. and Oklahoma.)

Families of Hemiptera

87 Third and fourth antennal segments equal or nearly equal in thickness to segment two. Fig. 340 88

Figure 340

Figure 340 *Ceratocapsus modestus* (Uhler)

Figure 341

Figure 341 *Orthotylus fuscicornis* Kngt.

87a Antennal segment two distinctly thicker than segments three and four (fig. 341) (sometimes only slightly thicker than segment three but always distinctly thicker than segment four) 90

88 Pronotum rather evenly narrowing laterally from posterior to anterior margins. Fig. 342 .... *Ceratocapsus*

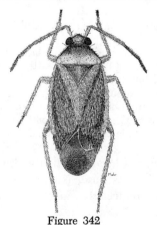

Figure 342

Figure 342 *Ceratocapsus pumilis* (Uhler)

This is a large genus of brown or reddish-brown to yellowish species whose members occur throughout the United States. Most if not all species are predaceous, but many appear to be associated with particular trees, shrubs or herbs. Fifty-six species have been described from north of Mexico.

88a Anterior half of pronotum evenly rounded and cylindrical posterior half of pronotum abruptly widened 89

89 Distal area of corium pruinose not polished; pronotal disc only slightly convex .... *Pamillia*

(Four scarce species occur in the southwestern United States. One has been taken in New Jersey.)

89a Distal area of corium and cuneus polished, strongly shining; basal half of pronotal disc strongly convex .... *Schaffneria*

(Five species of wide distribution occur from Michigan to Texas. All are rare ground-living forms.)

90 Pronotum finely but distinctly punctate 91

90a Pronotum smooth or rugose, never finely but definitely punctate 94

91 General coloration variegated, chiefly pale and orange coloration; eyes distinctly pedunculate and stalked .... *Labopella*

(A single species, *claripennis* Kngt., is known from north Mexico and Texas.)

91a General coloration blackish with paler areas of body confined to head, scutellum, embolium and posterior and lateral areas of pronotum; eyes not strongly pedunculate and stalked 92

92 Head completely black dorsally .... *Slaterocoris*

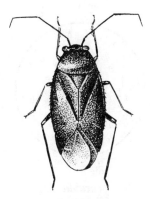

Figure 343

Figure 343 *Slaterocoris stygicus* (Say)

This is an ovoid jet-black shining mirid (about 4.5 mm) that is frequently abundant in old fields on goldenrod. It is widely distributed in eastern and central North America and at least in the north extends westward to Montana and California.

Forty-three species are known, all very similar in size and color. The identification of the different species often is very critical.

92a Head with at least lateral areas pale 93

93 Dorsal surface not shining, scutellum yellow .... *Lopidella*

(One species, *flavoscuta* Kngt., is known from Arizona.)

93a Dorsal surface strongly shining; scutellum black .... *Scalponotatus*

(Two species occur north of Mexico, in California and Arizona.)

94 Second antennal segment greatly enlarged and often flattened 95

94a Second antennal segment normally rounded, sometimes thickened but never greatly enlarged and flattened 97

95 First and second antennal segments with flattened hairs. Fig. 344 ...... *Heterotoma*

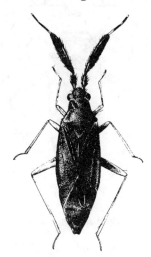

Figure 344

(One Palearctic species, *meriopterum* (Scopoli), (fig. 344) has been taken in New York and Washington.)

95a First and second antennal segments lacking flattened hairs 96

96 First antennal segment of relatively linear shape, not strongly enlarged on basal third .... *Ballella*

(One species, *basicornis* Kngt., occurs in Arizona.)

96a First antennal segment of male enlarged and swollen on basal third, tapering markedly to distal end .... *Daleapidea*

(Two species occur in the southwestern states.)

97 Pronotum raised posteriorly, projecting above scutellum; pleural areas of pronotum anteriorly separated from dorsal surface by a distinct suture .... *Semium*

Families of Hemiptera   175

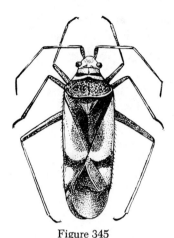

Figure 345

Figure 345 *Semium hirtum* Reuter

This is a bizarre little species (2.75 mm) with a velvety appearance throughout. The front wings are strikingly harlequined with white and brown coloration. The legs, antennae, head and sides of the pronotum are a deep rose-red color. It breeds on *Euphorbia* spp. and is known from coast to coast, although southern records are scarce.

A second species, *subglaber* Kngt., occurs in the southwest. Despite the arolia the genus is now placed in the Phylinae on the basis of the genitalia.

97a Pronotum not projecting over scutellum and without a distinct lateral suture 98

98 Dorsal surface clothed with scaly or flattened hairs intermixed with upright hairs. Figs. 346, 347, 348 .... 99 (see couplet 109)

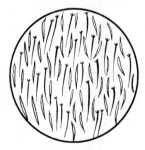

Figure 346

Figure 346 *Psallus* sp.

Figure 347

Figure 347 *Heterocordylus* sp.

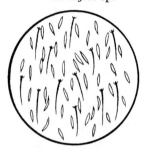

Figure 348

Figure 348 *Orthocephalus* sp.

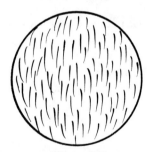

Figure 349

Figure 349 *Hoplomachus* sp.

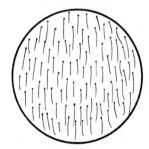

Figure 350

Figure 350 *Plagiognathus* sp.

98a Dorsal surface usually clothed only with a single type of pubescence (figs 349, 350), sometimes silky hairs intermixed, but these never strongly flattened and scalelike 117

99   Head with a sharp, well-defined basal margin, or, if questionable, then tylus elongate and extending far forward .... 100[6]

99a Head lacking a sharp, well-defined basal margin, or if somewhat well defined then hind femora with 3 or 4 longitudinal black lines 115

100 Tibiae with black or dark brown spots at bases of spines (at least on dorsal surface) 101

100a Tibiae lacking black or dark brown spots at bases of spines 104

101 First antennal segment rather thick, black, thickness about equal to twice thickness of fore tibia .... *Ceratopidea*

(One species, *daleae* Kngt., occurs in Nevada.)

101a First antennal segment less thickened, not twice as thick as fore tibia 102

102 Length of first antennal segment greater than width of vertex plus dorsal width of an eye .... *Squamocoris*

(Two species are known from Utah, Idaho, Nevada and Arizona.)

102a Length of first antennal segment not equal to width of vertex plus dorsal width of an eye. Fig. 351 103

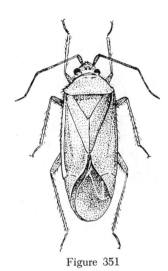

Figure 351

Figure 351 *Pseudopsallus* sp.

103 Body black, both sexes strongly brachypterous with short padlike wings .... *Anapus*

(See couplet 82)

103a Body not black, at least partially pale; wings fully developed ...... *Pseudopsallus*

(Three species occur in the western states.)

104 Second antennal segment moderately but distinctly thickened toward distal end 105

104a Second antennal segment slender not conspicuously thickened distally 106

105 Nonscalelike hairs on dorsal surface, thick, elongate, semierect, conspicuous. Fig. 348 .... *Orthocephalus*

---

6. In species of *Hesperocapsus* the posterior margin of the head is sharply defined only by vestiture. Males have the anterior margin of the genital cavity with 2 or 4 sclerotized plates or spines.

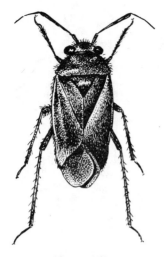

Figure 352

One introduced European species, *coriaceus* (Fabr.) (fig. 352), is sometimes common along roadsides in the northeastern states. A second species, *saltator* (Hahn), occurs in Alaska.

105a Nonscalelike hairs on dorsal surface sparse, very short and inconspicuous. Fig. 347 .... *Heterocordylus*

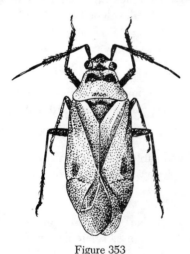

Figure 353

Figure 353 *Heterocordylus malinus* Reuter

This is a large black mirid (6.25 mm) sometimes entirely so but usually marked with red on the posterior half of the pronotum and with large areas (variable) on the front wings red. It lives on hawthorn and sometimes damages apple trees. The distribution is throughout the eastern states west to Minnesota and Missouri. A second species occurs in Texas.

106 Tarsal claws deeply split .... *Bifidungulus*

(Two scarce species occur in the western states.)

106a Tarsal claws not deeply divided. Fig. 279 107

107 Vertex very broad, three or nearly three times as great as width of an eye (females frequently short winged) 108

107a Width of vertex at most little more than twice width of an eye (females long winged) 109

108 Small to minute black species .... *Halticus*

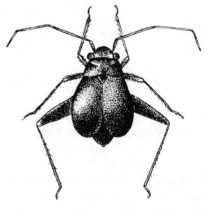

Figure 354

178 Families of Hemiptera

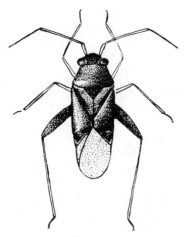

Figure 355

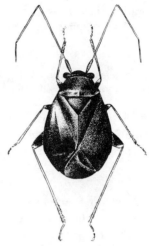

Figure 356

Figures 354 and 355 *Halticus bractatus* (Say)

This is a minute black insect (1.5-2.25 mm) with the first antennal segment, frequently the central part of the second and the base of the third antennal segments pale. Patches of silvery scale-like hairs are present on the front wing. These tiny mirids are known as Garden Flea Hoppers and jump actively when disturbed. They are frequently pests on various legumes and often can be collected on white clover in lawns. The sexes are strongly dimorphic, females having the wings entirely thickened and ovoid (fig. 354) so that they resemble small flea beetles, while the males have long normally constructed wings (fig. 355). The distribution is over the entire eastern and midwestern United States.

H. intermedius Uhler is a larger species (fig. 356) (3.5 mm) with yellow antennal segments, It occurs on *Clematis virginiana* from Pennsylvania to Colorado.

Figure 356 *Halticus intermedius* Uhler

One additional species, *apterus* (L.), that lacks scalelike pubescence on the hemelytra, occurs in Maine and eastern Canada.

108a Small to medium-sized greenish or pale colored species .... *Labopidea*

(Sixteen species occur which are mostly far western in distribution.)

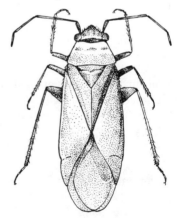

Figure 357

Labopidea allii Kngt. (fig. 357) is our commonest species and occurs on wild onion from the plains states east to Indiana.

Families of Hemiptera   179

109 Coloration dark brown; labium at most slightly surpassing mesosternum posteriorly .... *Noctuocoris*

(One scarce species, *fumidus* V. D., occurs from New York to Colorado across the northern states.)

109a Generally small greenish species; labium variable in length frequently considerably surpassing mesosternum 110

110 Length of first antennal segment less than width of head vertex, or if longer, black spots not present on pronotum 111

110a First antennal segment as long as or longer than width of head vertex, or if shorter, black scaly spots present on pronotum 113

111 Dorsal surface bearing two types of pubescence, consisting of simple pubescence intermixed with sericeous hairs, but without bristlelike hairs .... *Dichaetocoris*

(Twelve species occur in the western and southwestern states.)

111a Dorsal surface bearing at least two types of pubescence (figs. 346, 348), one type consisting of a few, or many, erect bristlelike hairs 112

112 Anterior margin of male genital cavity with from two to four sclerotized plates or spines (fig. 358); dorsal surface rather densely clothed with bristlelike pubescence .... *Hesperocapsus*

Figure 358

Figure 358 *Hesperocapsus artemisicola* Kngt.

(Fifteen species occur in the western and southwestern states.)

112a Anterior margin of male genital cavity lacking prominently projecting plates or spines; dorsal surface less densely ("sparsely") clothed with bristlelike pubescence .... *Melanotrichus*

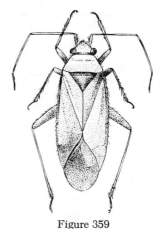

Figure 359

Figure 359 *Melanotrichus flavosparsus* Sahlb.

This is a small pale green mirid (4.0 mm) usually with a smoky colored wing membrane. The upper surface is clothed with dark upstanding hairs that are intermixed with silvery flat-lying hairs usually grouped into tiny patches. This small mirid often swarms in vast numbers on pigweed (*Chenopodium album*) and is found throughout the eastern and central states.

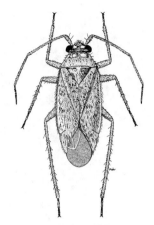

Figure 360

M. althaeae (Hussey) (fig. 360) is another green species similar in size but with the scaly hair patches black instead of silvery. It is known as the "Hollyhock Plant Bug" and is sometimes injurious to this plant in cultivated gardens. It is known from Illinois west to Colorado.

Twenty-eight additional species occur north of Mexico. Some are found in all parts of the country but the majority are western. (Some authors consider *Melanotrichus* to be a subgenus of *Orthotylus*).

113 Labium extending posteriorly beyond distal ends of hind coxae ... *Macrotyloides*

(Two species occur in the western states.)

113a Labium not extending posteriorly beyond hind coxae 114

114 Bristles black; pronotum lacking black scaly spots; first antennal segment much longer than width of vertex. Fig. 361 .... *Ilnacorella*

(Three species occur in the western states.)

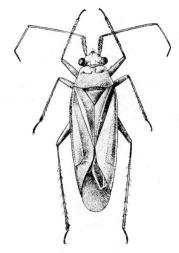

Figure 361

Figure 361 *Ilnacorella* sp.

114a Bristles light colored; pronotum always with black scaly spots; first antennal segment about equal in length to width of vertex .... *Ilnacora*

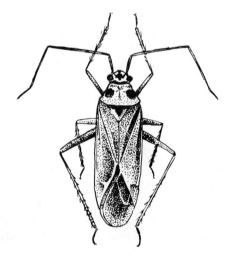

Figure 362

Figure 362 *Illnacora malina* (Uhler)

This is a beautifully marked black insect (5.5 mm) with the base of the pronotum and the front wings a rich green. The species breeds chiefly on goldenrod in damp places. It is often abundant throughout the eastern and central states west to Iowa and Missouri.

Thirteen additional species occur in America north of Mexico.

115 Second antennal segment swollen toward distal end; head rounded in front. Fig. 363 .... *Globiceps*

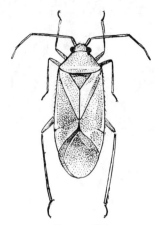

Figure 364

Figure 364 *Parthenicus* sp.

(There are sixty-eight described North American species, most of them small to minute in size. The majority of the species occur in the southwestern states.)

117 Eyes strongly pedunculate, set on prominent laterally projecting stalks .... *Labops*

Figure 363

Figure 363 *Globiceps* sp.

(Three scarce species occur in mountainous areas or at higher latitudes. The wings are often reduced and padlike.)

115a Second antennal segment evenly cylindrical; head usually bluntly pointed .. 116

116 Head strongly produced anteriorly, with a very large prominent tylus; body over 4 mm in length; hind femora not swollen with 3-4 black lines present .. *Argyrocoris*

(One rare species, *scurrilis* V. D., is found in Arizona.)

116a Head little produced anteriorly; body less than 4 mm in length; hind femora swollen lacking 3-4 black lines. Fig. 364 .... *Parthenicus*

Figure 365

Figure 365 *Labops hesperius* Uhler

This is a curious dark bluish-gray and white little insect with a very broad head and great protruding eyes. It sometimes occurs in large numbers in the western prairie country. The distribution is over the western states.

Seven additional species occur in North America. The distribution is chiefly northwestern but one species (*hirtus* Kngt.) also occurs in the mountains of New England and New York. In the western states members of the genus are sometimes destructive on range land.

117a Eyes usually sessile, at most projecting on very short head projections 118

118 Eyes rounded behind and placed near middle of head (in some species of *Diaphnidia* occupying nearly entire lateral head area) 119

118a Eyes almost straight behind and placed near to, or at, anterior margin of pronotum 125

119 Vertex of head sunken in (depressed) in center; wing membrane with one distinct cell present .... **Hyalochloria**

(One species, *caviceps* Reuter, has been taken in Florida.)

119a Vertex of head convex, not depressed in middle; wing membrane usually with two distinct cells present 120

120 First antennal segment marked with distinct black line on either side .... **Reuteria**

Reuteria irrorata (Say)

This is a small (4.25 mm) pale green to whitish insect with a very delicate habitus. It usually has darker greenish spots and blotches scattered about on the fore wings. It is found chiefly on elm trees in the northern states from New York west to Iowa. Five additional species occur in this country.

120a First antennal segment not marked with a pair of longitudinal black lines 121

121 Length of cuneus less than twice its basal width .... **Saileria**

(One species, *bella* (V. D.), occurs in California.)

121a Length of cuneus twice or more than twice width of cuneus at base 122

122 Eyes small, further removed from anterior margin of pronotum than diameter of first antennal segment. Fig. 366 .... **Paraproba**

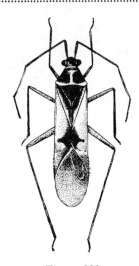

Figure 366

Figure 366 *Paraproba* sp.

(Five species occur in North America.)

122a Eyes large, placed posteriorly on head and not further removed from anterior pronotal margin than diameter of first antennal segment 123

123 Second antennal segment shorter than third .... **Brachynotocoris**

(One species, *delicatus* (Heid.), is known from eastern North America.)

123a Second antennal segment longer than third 124

124　Labium extending only to middle of mesosternum; posterior margin of pronotum convex at base .... *Diaphnidia*

(Two species occur in North America.)

124a　Labium reaching or exceeding mesocoxae; posterior margin of pronotum concave, sinuate or occasionally straight at base .... *Diaphnocoris*

Figure 368

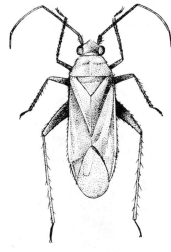

Figure 367

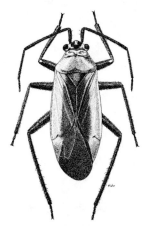

Figure 369

Figure 367　*Diaphnocoris pellucida* (Uhler)

This is a very slender, elongate, delicate pale greenish-white insect with the eyes and the tips of the tarsi contrasting dark brown to black. It has been taken on many trees and may be predaceous. This is a common and widely distributed insect throughout the eastern and central states.
　　Three additional species occur in North America.

125　Color generally strongly contrasting orange and black or red and black but sometimes gray and white or yellow; a well-defined suture on cheeks running from antennal socket to below eyes and frequently outlined by a dark stripe. Fig. 368. Fig. 369 .... *Lopidea*

Figure 369　*Lopidea confluenta* (Say)

This is a very large genus containing over 100 described species. Most of the species are remarkably similar in size, shape and color, but the male claspers take a great array of shapes and twists and their dissection is necessary for accurate specific identification. Some of our commonest species are *L. davisi* Kngt., which is frequently a phlox pest; *L. robiniae* (Uhler) which lives on black locust and *L. media* (Say) which is common on goldenrod. Despite their superficial similarity, definite "species groups" are present and their evolution and host adaptation offers an attractive field of study to the serious student.

125a　Without an oblique cheek suture running from antennal socket to beneath eye (or if present running into contact with lower eye margin); (color sometimes red and black) 126

126　Head vertex with a distinct raised ridge at posterior margin 127

126a Head lacking a raised ridge across posterior margin[7] 130

127 Posterior ridge (carina) on head vertex lacking a series of stout bristles. Fig. 370 .... *Orthotylus*

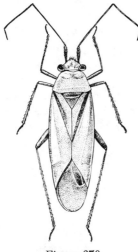

Figure 370

Figure 370   *Orthotylus* sp.

This is a large genus containing approximately forty-five North American species[8] many of which are common and widespread. The majority of the species are found on trees and shrubs and often have green and black markings. Much interesting work remains to be done on this complex and fascinating group of plant bugs.

127a Posterior ridge of head vertex bearing a series of stout bristles 128

128 First antennal segment nearly as long as width of head across eyes, always considerably longer than width of vertex .... *Blepharidopterus*

(One European species, *angulatus* (Fallen), has been reported from Alberta, British Columbia and Nova Scotia.)

128a First antennal segment at most only as long as width of head vertex 129

129 Color yellowish green; pronotum not having lateral margins carinate .... *Labopidea*
(See couplet 108a.)

129a Color dark, usually dark gray or blue and often with traces of reddish; lateral pronotal margins usually laminate .... *Hadronema*

Figure 371

Figure 371   *Hadronema militaris* Uhler

This is a grayish-blue insect with white lateral wing margins that has the general aspect of a stout dark *Lopidea*. It is a common insect on prairie grasses in the western states but with a challenging distribution pattern (prairie peninsula) as it has also been reported from New York, Michigan and Indiana. Eleven additional species are found in the western United States.

---

7. In some species of *Labopidea* the posterior ridge is relatively weakly developed but these are greenish species which is not true of species keying to 127.
8. *O. virescens* D. & S. and *O. concolor* (Kirsh.), both Palearctic species, have recently been taken on broom (*Sarothamnus scoparius* [L.]) in California.

130   Pronotum with a narrow anterior collar. Fig. 372 .... *Mecomma*

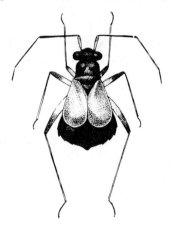

Figure 372

Figure 372   *Mecomma* sp.

(Four scarce species occur in North America, usually in mountainous regions or in higher latitudes.)

130a   Pronotum without an anterior collar .... *Cyrtorhinus*

(One species, *caricis* (Fallen), occurs in the northern states south to Virginia.)

131   Pseudarolia absent (fig. 278); pronotal collar usually present, if absent then claws very long, smooth and slender 132

131a   Pseudarolia present (fig. 279); sometimes minute and difficult to see, in which case pronotal collar is absent and claws not unusually long and slender 146

132   Claws smooth at base, long and slender. Figs. 278, 279 133

Figure 373

Figure 373   *Brachyceratocoris nevadensis* Kngt.

132a   Claws toothed, cleft or distinctly thickened at base. Fig. 373 135

133   Head short and strongly declivent; eyes globose rising a considerable distance above dorsal surface of head .... *Cylapus*

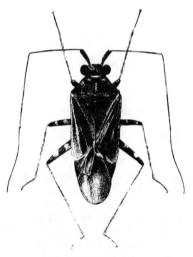

Figure 374

Figure 374   *Cylapus tenuicornis* Say

This is a handsome brownish insect marked with white on the scutellum, middle of corium, apex of clavus, base of cuneus and adjacent to cell of wing membrane. The hind femora have two conspicuous white bands and the hind tibiae have a broad white median annulus. The eyes are very large and protruding and the antennae are extremely long and sweeping. This is a rare species that is associated with fungi on tree trunks. The genus is apparently a very old one with species scattered over much of the world, but with only a single species in North America where it has been taken in the northeastern and north central states.

133a Head long, slender and pointed, not declivent; eyes not produced prominently above dorsal surface of head 134

134 Pronotal collar present. Fig. 375 .... *Fulvius*

Figure 375

Figure 375   *Fulvius* sp.

Two species are widely distributed north of Mexico. The members of this genus look superficially like long-legged Anthocoridae and are usually found on the ground or on the bark of fallen trees. They are predaceous.

134a Pronotal collar absent .... *Peritropis*

This genus contains two rare North American species, both of which are grayish brown with numerous large white spots sprinkled over the dorsal body surface (fig. 376). Little is known of their habits but they appear to be bark inhabiting species. The scattered records indicate a wide range.

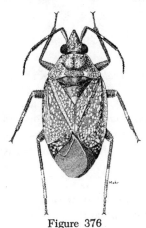

Figure 376

Figure 376   *Peritropis husseyi* Kngt.

135 Head elongate and pointed, nearly as long as pronotum; antennae very short 136

135a Head strongly bent over (declivent), frequently at nearly a right angle to vertex; antennae not appreciably shorter than in most other plant bugs 137

136 First antennal segment at most reaching apex of head; second antennal segment enlarged and flattened. Fig. 377 .... *Hesperophylum*

Figure 377

Figure 377   *Hesperophylum arizonae* Kngt.

(One very rare species, *heidemanni* Reuter & Poppius, is known from such widely separated localities as Arizona, Washington, D. C., Iowa and New Hampshire. A second species occurs in Arizona.)

136a First antennal segment exceeding apex of head; second antennal segment not strongly flattened ...... *Conocephalocoris*

(One species, *nasicus* Kngt., is known from Arizona.)

137 Pronotum with a groove or impressed line running from anterior corner to behind calli. Fig. 378 138

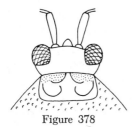

Figure 378

Figure 378 *Guanabarea* sp.

137a Pronotum lacking an impressed line from anterior corner to posterior margin of calli **140**

138 Anterior portion of pronotum swollen, extending as a small hood over base of head .... ***Clivinema***

(Ten species are found in the western states.)

138a Anterior pronotal margin not extending hoodlike over base of head **139**

139 Lateral pronotal margins distinctly ridged (carinate) .... ***Largidea***

(Ten species occur north of Mexico of which nine are western and one is found in New York.)

139a Lateral margins of pronotum not carinate. Fig. 379 .... ***Bothynotus***

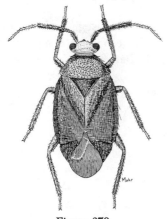

Figure 379

Figure 379 *Bothynotus modestus* Wirtner

(Four scarce species occur in North America and have an extensive but poorly understood distribution.)

140 Front wings almost completely transparent and glassy in appearance .. ***Hyaloides***

Hyaloides vitripennis (Say)

This is a beautiful "glassy" insect (4.75 mm) with red or black margins to the scutellum and the inner and apical margins of the corium. The color markings on the pronotum and antennae are variable. The head is globular with the eyes forming a smooth line with the curvature of the head. It is widely distributed in the eastern and central states where it is most frequently taken on grape vines and feeds upon aphids. Two additional species occur in North America.

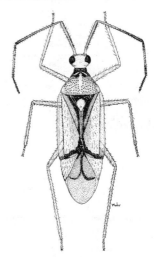

Figure 380

Figure 380 *Hyaloides harti* Kngt.

140a Fore wings not glassy and transparent, but subopaque throughout; (if wings appear somewhat "glassy" then antennal segment 2 abruptly clavate distally) .. **141**

141 Vertex of head with a distinct groove present (sulcate) (slight in some species of *Eustictus*); frons striated **142**

141a Vertex of head not grooved; frons usually polished and not or very faintly striated 143

142 Second antennal segment strongly and abruptly enlarged toward distal end (clavate); third and fourth antennal segments thick, somewhat enlarged in middle (fusiform). Fig. 381 .... *Diplozona*

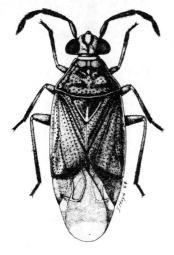

Figure 381

(One species, *collaris* V. D. (fig. 381) occurs in Florida.)

142a Second antennal segment at most only slightly enlarged distally, third and fourth segments linear throughout. Fig. 382 .... *Eustictus*

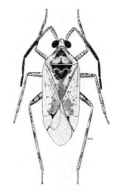

Figure 382

Figure 382 *Eustictus salicicola* Kngt.

Twenty-one species of this genus occur in this country. The genus is widespread although most of the species are southern and southwestern in distribution. They are large (6-8 mm) plant bugs, not usually very common, and apparently predaceous on small arthropods.

143 Tylus extending forward beyond distal end of first antennal segment; embolium of front wing very broad. Fig. 383 .... *Eurychilopterella*

Figure 383

Figure 383 *Eurychilopterella luridula* Reuter

(Three scarce species occur in North America.)

143a Tylus not extending forward beyond distal end of first antennal segment; embolium not unusually broadened 144

Families of Hemiptera   189

144   Second antennal segment usually nearly linear, at most only moderately enlarged toward distal end. Fig. 384 .... *Deraeocoris*

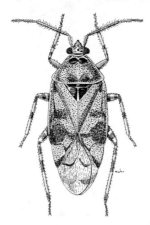

Figure 384

Figure 384   *Deraeocoris aphidiphagus* Kngt.

Deraeocoris nebulosus (Uhler)

This is a small member of the genus (3.5-4 mm) and is of an olive-yellow color with black markings scattered over the upper surface. The membrane is clear with two small brown spots present on the outer half. It occurs most commonly on oak and is predaceous. The distribution is throughout most of the United States west to Texas and Colorado. It is often a very common species.

   D. histrio (Reuter) is a bright red species, somewhat larger than *nebulosus* with a pair of large black spots on the corium. It is predaceous and usually found on smartweed over the northern half of the United States.

   This is a very large genus containing fifty-six species north of Mexico. Many are widely distributed and reasonably common. Much biological investigation is needed of members of this genus for, while the species appear to be predaceous, they are often apparently confined to certain plants. Interesting problems concerning the evolution of predatory habits certainly await the careful investigator.

144a   Second antennal segment distinctly and strongly enlarged toward distal end (clavate) 145

145   Small insects little more than 4 mm in length; cuneus of front wings strongly bent downward; a row of punctures present along claval suture .... *Klopicoris*

(One species, *phorodendronae* V. D., occurs in California.)

145a   Large species (8 mm); cuneus not strongly bent downward; claval suture lacking a row of punctures .... *Deraeocapsus*[9]

(Two species occur in California.)

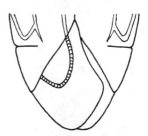

Figure 385

146   Membrane of front wing with 1 cell present (fig. 385) (sometimes membrane lacking); tarsi thickened toward apices; pseudarolia arising from ventral surface of claw. Fig. 386 147

Figure 386

Figure 386   *Pycnoderes* sp.

146a   Membrane of front wing with 2 cells present (fig. 387) (distal cell sometimes very small); tarsi linear; pseudarolia arising from base or inner margin of claw 154

9. *Strobilocapsus* Bliven may key near here.

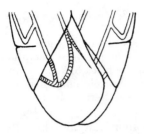

Figure 387

147 Fore wings beetlelike, lacking a membrane and without a differentiated clavus and corium .... *Hemisphaerodella*

(One species, *mirabilis* Reuter, occurs in Florida.)

147a Fore wings having a membrane at least in male sex, usually in both sexes; clavus and corium usually well differentiated **148**

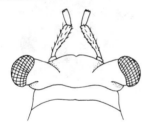

Figure 388

Figure 388 *Hesperolabops* sp.

148 Eyes strongly stalked, stalk at least as long as width of an eye. Fig. 388 .... *Hesperolabops*

(One species, *gelastops* Kirk., of this curious genus occurs on cacti in Texas.)

148a Eyes sometimes pedunculate but not set on distinctly elongate stalks that are as long as eye width **149**

149 Labium extending posteriorly to or beyond hind coxae **150**

149a Labium not extending posteriorly to hind coxae **151**

150 Pronotal calli strongly swollen and convex .... *Caulotops*

(Two scarce species occur in Arizona.)

150a Pronotal calli not strongly swollen and convex .... *Halticotoma*

(Five species occur chiefly in the southern and southwestern states on yucca.)

151 Pronotal collar distinct but not wider than width of second antennal segment .. .. *Monalocoris*

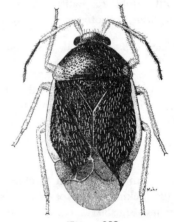

Figure 389

Figure 389 *Monalocoris americanus*
Wagner & Slater

This is a very small (2.5 mm) dark brown to blackish oval plant bug with pale yellowish legs and antennae. The third and fourth antennal segments are dark and the embolium in most specimens is pale yellow. This tiny insect is often common on ferns in shaded woodlands. It is found over the entire eastern and midwestern United States and in south Canada.

A second species occurs in Florida.

Families of Hemiptera 191

151a Pronotal collar either indistinct or distinct, but if distinct then its width greater than that of second antennal segment 152

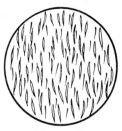

Figure 390

Figure 390 *Atractotomus* sp.

152 Fore wings bearing silvery, silky or woolly hairs (figs. 346, 390); width across eyes about equal to width across anterior margin of pronotum .... *Cyrtocapsus*

(One species, *caligineus* Stål, has been taken in California and Florida.)

152a Fore wings not bearing silky or woolly hairs; width across eyes greater than that across anterior margin of pronotum by at least one-half width of an eye 153

153 Embolium narrow and thickened; front wings lacking areas of silvery spots or patches. Fig. 391 .... *Sixeonotus*

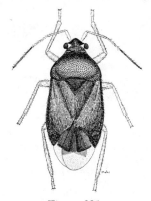

Figure 391

Figure 391 *Sixeonotus insignis* Reuter

(Seventeen species of this genus occur north of Mexico. The majority are southern in distribution. They are usually black to dark brown in color with the pronotum coarsely punctured and moderately inflated on the posterior lobe.)

153a Embolium flat and broadly expanded; front wings with areas of silvery spots or patches. Fig. 392 .... *Pycnoderes*

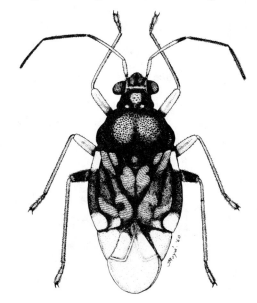

Figure 392

Figure 392 *Pycnoderes quadrimaculata* (Guerin)

(Ten chiefly southern species are known. The antennae are long and sweeping. The pronotum is very strongly swollen on the posterior lobe and usually has a median depression so that a pair of swollen lobes are produced.)

154 Pronotum with a well-developed anterior collar; or if collar is lacking species are ant mimics 155

154a Pronotum lacking an anterior collar; never ant mimics 162

155 Ant mimics with abdomen strongly constricted at base (fig. 293); claws long, slender, not bent at base 156

155a Not ant-mimetic species, abdomen not strongly constricted at base; claws when long and slender basally bent 160

156 Pseudarolia large, reaching or nearly reaching apices of claws. Fig. 279 157

156a Pseudarolia minute, not nearly reaching apices of claws. Fig. 393 159

Figure 393

157 Second antennal segment strongly thickened toward distal end (clavate); labium extending backward to or almost to hind coxae; females with fully developed front wings .... *Teleorhinus*

(Seven species occur in the United States.)

157a Second antennal segment at most very slightly thickened toward distal end, labium extending backward only to middle coxae; females with short reduced front wings or wingless 158

158 Pseudarolia free from claw toward tip (fig. 279); females wingless; second antennal segment linear. Fig. 394 ....
.... *Coquillettia*

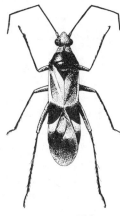

Figure 394

Figure 394 *Coquillettia* sp.

(Twenty-three species of these wonderful, chiefly western, ant mimics occur in North America. In the northern states they are found only a little east of the Mississippi River but in the south extend to North Carolina and Florida. They occur on the ground, usually in open country and often associated with ants, although there is no evidence as yet of an ecological interaction between them. The wingless females are such remarkable mimics as to "fool" even an experienced collector.)

Figure 395

158a Pseudarolia completely joined to claw (fig. 395); females have short wings; second antennal segment slightly thickened toward distal end .... *Orectoderus*

(Four species occur, three western and one scarce but widely distributed over much of the United States.)

159 Scutellum of females bearing an upright spinelike projection ...... *Heidemanniella*

(One species, *scutellaris* Poppius, occurs in Arizona.)

159a Scutellum of females lacking an upright spinelike projection .... *Closterocoris*[10]

(One species, *amoenus* (Provancher), occurs from California east to Kansas and Colorado.)

160 Eyes large, removed from anterior margin of pronotum by a distance less than half lateral width of an eye. Figs. 396 and 397 .... *Cyrtopeltis*

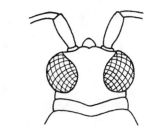

Figure 396

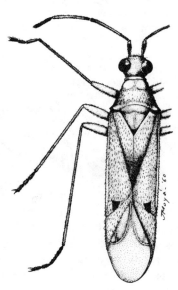

Figure 397

Figure 397 *Cyrtopeltis tenuis* Reuter

(Six North American species occur in the western and southern states.)

160a Eyes removed from anterior margin of pronotum by a distance equal to or greater than half lateral width of an eye. Figs. 398, 399 161

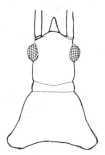

Figure 398

Figure 398 *Macrolophus brevicornis* Kngt.

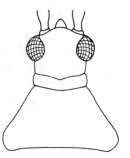

Figure 399

Figure 399 *Dicyphus agilis* (Uhler)

161 Eyes very small (fig. 398), usually separated from anterior margin of pronotum by a distance equal to or greater than length of an eye seen from above; head about as long as wide; head and pronotum almost entirely greenish yellow; pronotum lacking an arcuate furrow across middle .... *Macrolophus*

(Four species occur in North America, two of which are widely distributed.)

10. Knight (1968) keys this genus in the Mirini and the male genitalia support this placement.

161a Eyes relatively large (fig. 399), separated from anterior margin of pronotum by a distance less than length of an eye seen from above; head slightly wider than long; pronotum often with a deep arcuate furrow across middle, if absent then head mostly black and pronotum with brown or black at least laterally .... *Dicyphus*

Dicyphus cucurbitaceus Spinola

This is a small (3.5-4.5 mm) slender parallel-sided insect, yellow in color with a strongly contrasting black head, thorax and second antennal segment. The wings often have reddish markings. It is most often taken on raspberries but it also occurs on many other plants. The distribution is throughout much of the country, south in the east to Virginia.

Twenty additional species occur north of Mexico, several of them widely distributed.

162 Second and third antennal segments strongly and conspicuously inflated, bearing erect flat black scalelike hairs. Fig. 400 163

Figure 400

Figure 400 *Beamerella personatus* Kngt.

162a Second segment sometimes strongly inflated, but third segment always relatively slender and cylindrical, lacking erect flat black hairs 164

163 Tylus and head laterally shining black, giving impression of a face mask, fourth antennal segment inflated .... *Beamerella*

(One species, *personatus* Kngt., is known from Texas.)

163a Tylus and sides of head not shining black; fourth antennal segment tapering, not conspicuously inflated .. *Larinocerus*

(One species, *balius* Froeschner, occurs in California and Nevada.)

164 Pronotum distinctly punctured with lateral margins expanded, flattened and somewhat reflexed .... *Pronotocrepis*

(Three species occur in the western states.)

164a Pronotum impunctate or at most very finely punctate, lateral margins not expanded and reflexed 165

165 Body surface bearing scalelike hairs or flattened silvery hairs. Figs. 346, 347, 348, 390 166

165a Body lacking scalelike or flattened silvery hairs (figs. 349, 350) (in cases where somewhat flattened hairs appear to be present among other hairs then tibial spines always lack black spots at their bases) 184

166 Front of head bent nearly at right angles to long axis of body (fig. 401) very little produced in front of antennal bases when viewed from above. Fig. 402 167

Figure 401

Figure 401 *Ankylotylus pallipes* Kngt.

Families of Hemiptera 195

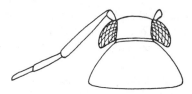

Figure 402

166a Head in dorsal view noticeably produced in front of bases of antennae. Fig. 403 171

Figure 403

Figure 403 *Atractotomus mali* (Mey.-Dur)

167 Head very wide, width equal to or slightly greater than width across base of pronotum .... *Beckocoris*

(One species, *laticephalus* Kngt., is known from Nevada.)

167a Head width not equal to basal width of pronotum 168

168 Distance between buccula and lower margin of eye not greater than thickness of first antennal segment .... *Rhinacloa*

(Four small blackish or brown species with distally thickened second antennal segments belong here and are widely distributed.)

168a Distance between buccula and lower margin of eye greater than thickness of first antennal segment 169

169 Tibial spines usually lacking a black spot at base, if present tylus not curved horizontally backward at distal end nor sharply bent at middle, or hemelytra and tibae not orange-red .... *Lepidopsallus*

Lepidopsallus rubidus (Uhler)

This is a small (3.25-3.5 mm) reddish-brown to nearly black species with numerous black spines on the tibiae. The body surface is densely covered with pale yellow scalelike hairs although a variety *atricolor* Kngt. has been described with silvery hairs. (This variety appears to be without geographic significance.) This species is often common on willows and is distributed over most of the United States.

Seventeen additional North American species are known, several of which are widely distributed. *Maurodactylus* (not in key) *consors* Uhler and *semiustus*. V. D. may belong to this genus.

169a Either with tylus sharply bent at middle and distal end bent horizontally backward (fig. 401), or hemelytra and tibiae orange-red; tibial spines with a black spot at base or spines and tibiae uniformly black 170

170 Tibial spines pallid to yellowish, femora never black .... *Ankylotylus*

(One species, *pallipes* Kngt., occurs in Nevada.)

170a Femora and tibial spines black .... *Merinocapsus*

(One species, *ephedrae* Kngt., occurs in Nevada.)

171 Tylus sharply produced with apex pointed, not noticeably downcurved at distal end when viewed laterally 172

171a Tylus somewhat vertical, conspicuously downcurved when viewed laterally .. 173

(In *Hoplomachidea* the head appears to us sharply produced but all existing keys treat it as blunt and we carry it there provisionally. Our only species is a rather large insect with brown and yellow variegated coloration and with a pale yellow-brown median stripe running from the base of the tylus to the apex of the scutellum.)

172 Second antennal segment only 1/2 or less than 1/2 width of pronotum across base; cuneus with pale basal and apical areas; both sexes with second antennal segment strongly thickened .... *Ranzovius*

(One species, *moerens* Reuter, is reported from California, Texas, Arizona and Florida.)

172a Second antennal segment as long as, or nearly as long as maximum pronotal width, always much more than 1/2 width; cuneus uniformly dark; males only with second antennal segment strongly thickened .... *Criocoris*

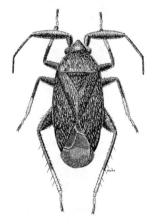

Figure 404

Figure 404 *Criocoris saliens* (Reuter)

This is a small (2.75-3 mm) black species which has the body sprinkled with white scalelike hairs. In the males the first two antennal segments are strongly thickened but in the females only the first segment is thick. It is often common on bedstraw (*Galium*) and occurs nearly throughout the country.
  A second species, *canadensis* V. D., occurs in Quebec and Ontario.

173 Second antennal segment strongly thickened (less so in females), much broader than first segment .... *Atractotomus*

(Thirteen North American species are recognized, most of them confined to the western states.)

173a Second antennal segment not strongly thickened, usually more slender than segment one 174

174 Pseudarolia large, attached only at basal angles and extending free and parallel with claws to apex. Fig. 279 175

174a Pseudarolia shorter, sometimes minute or lacking. Fig. 393 176

175 Rows of conspicuous large black spots present on hind femora ...... *Tannerocoris*

(One species, *sarcobati* Kngt., is known from Washington, Colorado, Idaho, Utah and South Dakota.)

175a Hind femora lacking rows of large, conspicuous black spots .... *Nevadocoris*

(Three species are known from Nevada.)

176 Tibial spines without a black spot at base (although tibiae sometimes spotted); color of body never black with white flattened scalelike hairs 177

176a Tibial spines with a black spot at base or spines and tibiae uniformly black; or black species with white scalelike hairs present. Figs. 346, 347, 390 178

Families of Hemiptera 197

177   Hemelytra spotted, membrane irregularly sprinkled with dark spots and blotches (conspurcate); tibiae spotted .. *Keltonia*

(Five species occur north of Mexico.)

177a   Hemelytra including membrane not sprinkled with dark spots; tibiae dark, unspotted .... *Reuteroscopus*

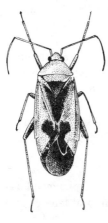

Figure 405

Figure 405   *Reuteroscopus ornatus* (Reuter)

This is a small (3.5 mm) yellowish-green species usually very easily recognized by the broad black band across the apical portion of the corium. The scutellum, clavus and membrane are also dark brown to black. This is often a very abundant insect on ragweed everywhere east of the Great Plains.
        (Fourteen additional species occur north of Mexico.)

178   Wing membrane markings conspurcate; male genital segment with tubercle present on left side (labium not attaining middle of abdomen) .... *Phymatopsallus*

(Twenty species occur in the southwestern states.)

178a   Wing membrane not conspurcate (or if somewhat so labium reaches middle of abdomen) 179

179   Femora and tibiae with large brownish-black spots; sexes dimorphic, females brachypterous with abdomen broad and subtriangular .... *Hoplomachidea*

(One species, *consors* (Uhler), occurs in Nevada and California. See note at couplet 171a).

179a   Lacking large brownish-black leg spots; sexes not dimorphic 180

180   Pseudarolia absent or very small (fig. 277); claws "broadly curved" .... *Megalopsallus*

(Six or seven species occur in the western states.)

Figure 406

Figure 406   *Microphylidea prosopidis* Kngt.

180a   Pseudarolia present and evident (fig. 406); claws "more sharply curved" 181

181   Antennal segment two with distinct black spots; sericeous hemelytral pubescence grouped into small spots .... *Pseudatomoscelis*

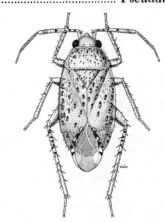

Figure 407

Figure 407   *Pseudatomoscelis seriatus* (Reuter)

Pseudatomoscelis seriatus is easily recognized by its pale yellowish color speckled with brownish-black spots. The second antennal segment has four to five very conspicuous black spots or bands.

This species is known as the "Cotton Flea Hopper" and is often a seriously destructive insect to the buds of cultivated cotton, although it also occurs on other plants. It is distributed throughout the southern and southwestern states.

181a Antennal segment two without conspicuous black spots; sericeous hemelytral pubescence not grouped into spots. Fig. 346 182

182 Head broad, vertical, tylus not visible when insect viewed from above; width of vertex equal to or greater than half width of head. Fig. 408 .... *Europiella*

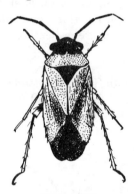

Figure 408

Figure 408   *Europiella* sp.

(Thirty-seven species occur in North America chiefly west of the Mississippi River.)

182a Head not so broad, width of vertex usually not equal to half width of head, if so then tylus is visible from above 183

183 Antennal segment two as long as or longer than basal width of pronotum .... *Psallus*

This is a large genus containing thirty-four North American species. Many of them occur on trees as well as herbaceous plants. Identification is often very critical as species are closely related and much work remains to be accomplished in the classification and biology of the genus.

183a Antennal segment two shorter than basal width of pronotum .... *Sthenarus*

(Five species occur in the southern or southwestern states.)

184 Second antennal segment usually subequal to or shorter than width of head across eyes, if longer then hind femora are light with dark spots at least on anterior aspect 185

184a Second antennal segment usually longer than width of head across eyes, when subequal hind femora not light with dark spots 193

185 Usually light-colored species with femora light-colored and having large black spots 186

185a Generally dark-colored species; if light colored then femora not having conspicuous black spots but sometimes completely black or dark brown 190

186 Distance between eyes less than 1 1/2 times diameter of 1 eye as seen from above; tibial spines on hind legs obscured by long hairs .... *Brachyceratocoris*

(One species, *nevadensis* Kngt., occurs in Nevada.)

186a Distance between eyes more than 1 1/2 times diameter of 1 eye as seen from above; tibial spines on hind legs prominent 187

187   Cuneus with large black spot, calli dark; hairs of dorsum of body black, more or less erect; size over 3 mm .... *Phyllopidea*

(Four species occur in the western states.)

187a Cuneus lacking large black spot, pronotum more or less uniformly colored; body hairs not erect; size less than 3 mm 188

188   Distance from lower margin of eye to buccula about equal to height of eye .... *Atomoscelis*

(One species, *modestus* (V. D.), occurs in the Great Plains and western states.)

188a Distance from lower margin of eye to buccula equal to or less than 1/2 height of an eye 189

189   Body length 2.5-3 mm .... *Campylomma*

Campylomma verbasci (Meyer)
This is a very small (2.5-3 mm) inconspicuous pale yellowish plant bug. It is marked with black on the distal half of the first and proximal area of the second antennal segments. The femora and tibiae of the middle and hind legs have conspicuous black spots. This is an introduced European species that is often very common on mullein in the northern states.

189a Body length barely exceeding 1 mm (1.02-1.03) .... *Mineocapsus*

(One species, *minimus* Kngt., is known from Utah.)

190   Body hairs very short so that insects appear nearly hairless above (glabrous) and of dull texture; tibial spines short, becoming obsolete on proximal 1/3 of tibiae .... *Conostethus*

(One species, *americanus* Kngt., is known from Nebraska, Colorado, Montana and South Dakota.)

190a Body hairs longer and distinct to give upper surface a definitely hairy and usually shining appearance; tibial spines relatively elongate, placed throughout tibiae 191

191   Hind femora pale without black or dark brown spots .... *Microphylidea*

(Two species occur. The distribution is Arizona, Utah and Nevada.)

191a Hind femora either black, dark brown or pale, but if pale then some black or dark brown spots are present 192

192   Males with first and second antennal segments greatly thickened (fig. 409); hemelytra black or grayish black with a pale mark on clavus .... *Spanagonicus*

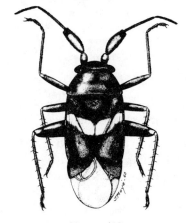

Figure 409

Figure 409   *Spanagonicus albofasciatus* Reuter

This is a very small dull-colored insect seldom over 2.5 mm long. The males are easily recognizable by the greatly swollen first and second antennal segments, females however are less distinctive. The color is somewhat variable, often with pale areas on the wings in addition to the claval mark. This is often a very common insect in the southern and southwestern states but occurs as far north as New York in the East. It has been reported injuring golf greens and is apparently a grass feeder.

192a First and second antennal segments of male relatively slender, similar to those of females (if with somewhat thickened antennal segments then clavus lacks a pale mark) .... *Chlamydatus*

Chlamydatus associatus (Uhler)

This is a small (2.5 mm) chiefly black species with strongly contrasting yellow legs, with the exception of the dark hind femora. It is a very abundant species on ragweed and many other plants throughout much of North America.

Twenty-one additional species occur north of Mexico.

193   Pseudarolia large, reaching to, almost reaching to, or projecting slightly beyond apices of claws. Fig. 279 194

193a Pseudarolia minute or invisible (fig. 278), if evident never approaching apices of claws 197

194   Eyes almost touching anterior margin of pronotum .... *Amblytylus*

Amblytylus nasutus (Kirschbaum)

This is a rather small (3-3.75 mm) nearly uniformly pale yellowish-green plant bug, ovoid in shape with a rather pointed head. The females have the wings shortened and not reaching the end of the abdomen. *A. nasutus* is a European species that has been introduced into North America. It appears to be rapidly increasing. It sometimes swarms in late spring on Kentucky Blue Grass and related grasses in meadows and pastures. It is now known from the northeastern states, west to Indiana and Michigan and south to Kentucky, but probably will continue to extend its range for some time.

194a Eyes clearly removed from anterior margin of pronotum 195

195   Dorsum with scattered black bristles or hairs .... *Macrotylus*

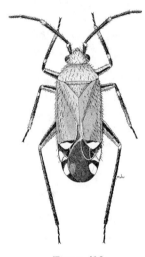

Figure 410

Figure 410   *Macrotylus amoenus* Reuter

This is a beautifully colored tiny mirid (2-2.5 mm) of a bright green color with a dark membrane and a series of striking large white spots laterally on the cuneus and membrane. It is sometimes common in the east on aster, and has also been taken in Illinois and Iowa.

Thirteen additional species occur in North America, all but one restricted to the far western states.

195a Body with light colored vestiture 196

196   Hemelytra including cuneus uniformly straw-yellow in color; cuneus much longer than scutellum .... *Lopus*

Lopus decolor (Fallen)

This is a medium-sized species (4.5 mm) of a dull straw-yellow color and usually with three dull brown bands on the posterior part of the pronotum. It apparently is an introduced European species and on the eastern seaboard sometimes swarms on sedges. *L. decolor* occurs in the east from Maine to Virginia and has also been reported from California.

196a Hemelytra mostly dark brown with cuneus and margins, lighter; cuneus not much longer than scutellum ...... **Nicholia**

(One species, *eriogoni* Kngt., occurs in Arizona.)

197 Margin of eye well separated from antennal socket, minimum space between them more than 1/3 as great as diameter of socket (fig. 411); margin of eye near antennal socket nearly evenly rounded 198

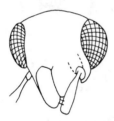

Figure 411

Figure 411 *Megalocoleus molliculus* (Fallen)

197a Margin of eye touching or almost touching antennal socket, minimum space between not more than 1/8 as great as diameter of socket (fig. 412); margin of eye usually somewhat "curved in" and irregular in outline (emarginate) near antennal socket 200

Figure 412

Figure 412 *Plagiognathus* sp.

198 First antennal segment dark or partially dark .... **Monosynamma**

(One species, *bohemani* (Fallen), is distributed throughout the northern states on willows and is often locally common.)

198a First antennal segment light 199

199 Labium extending backward onto third abdominal segment .... **Megalocoleus**

(One European species, *molliculus* (Fallen), has been taken in Massachusetts.)

199a Labium not extending backward beyond hind coxae .... **Oncotylus**

(One species, *guttulatus* Uhler, occurs in western North America.)

200 Labium only slightly extended onto mesosternum, at most just reaching middle coxae .... **Myochroocoris**

(One species, *griseolus* Reuter, occurs in Texas.)

200a Labium extending to or beyond hind coxae 201

201 Body hairs entirely composed of black setae-like hairs (fig. 349), always lacking a clothing of silky or woolly hairs .... **Hoplomachus**

(One species, *affiguratus* (Uhler), occurs in the northern and northwestern states.)

201a Body bearing silky or woolly hairs, sometimes, but not always, intermixed with setae-like hairs (fig. 346) (if black setae-like hairs present (*Chaetophylidea*) then second antennal segment longer and more slender than segment one) 202

202 Hind tibiae with black spines but these always lacking dark spots at their bases 203

202a Hind tibiae with either light yellow, colorless or dark spines but when spines are dark then black spots are present at their bases (occasionally these spots are obscure) 206

203 General coloration greenish yellow .... *Tuponia*

(Three species are known to occur north of Mexico, two in California and one in Colorado.)

203a General coloration dark, at least with red-brown or black areas on head and pronotum (but see *Microphylellus*) .. 204

204 Second antennal segment slightly swollen at outer end to become as wide as segment one; general color dark red .... *Rhinocapsus*

Figure 413

Figure 413 *Rhinocapsus vanduzeei* Uhler

This is a dark red mirid (3.5-4 mm) with the second antennal segment usually reddish yellow on the basal third and dark on the outer portion but sometimes completely darkened. The wings are dark brown to nearly black and the legs dull yellowish. It is found on wild raspberry in the eastern states west to Michigan and south to North Carolina.

Two additional species occur in the eastern states.

204a Second antennal segment evenly linear throughout, not as wide as segment one; color black or rarely pale, rarely mixed with reddish 205

205 Pronotum campanulate, strongly flaring laterad at humeral angles. Fig. 414 .... *Tytthus*

Figure 414

Figure 414 *Tytthus* sp.

(Seven generally scarce species are widely distributed in North America and, except for one northeastern species, are southern and western in distribution.)

205a Pronotum conventionally shaped, not prominently campanulate .... *Microphylellus*

(A genus of small, frequently black, species [many widely distributed] and containing eighteen species north of Mexico.)

206 Hind tibiae with dark or black spines, usually with dark spots at their bases 207

206a Hind tibiae with yellow or colorless spines 208

207 Head and pronotum bearing erect black bristles .... *Chaetophylidea*

(One species, *moerens* (Reuter), occurs chiefly in the western states.)

207a Head and pronotum clothed with decumbent hairs or pubescence, without erect bristles. Fig. 415 .... *Plagiognathus*

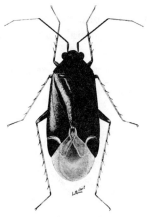

Figure 415

Figure 415 *Plagiognathus obscurus* Uhler

Plagiognathus politus Uhler

A small (3.5-4 mm) black plant bug of a shining black color covered with whitish hairs. The femora are black with pale tips and the tibiae yellowish white. The second generation is lighter in color and yellowish on the pronotum and scutellum. This is one of the most common species of mirids and occurs everywhere east of the Great Plains on ragweed, goldenrod and other Compositae.

This is a very large genus containing an additional seventy-seven species north of Mexico. Many of them are very closely related.

208 Third segment of hind tarsus as long as first and second segments combined .... *Plesiodema*

(One species, *sericea* (Heid.), is found on basswood from Washington, D.C. west to Missouri and Iowa and north to Michigan and Minnesota.)

208a Third segment of hind tarsus shorter than segments 1 and 2 combined .... *Icodema*

(One species, *nigrolineata* (Kngt.), occurs on burr oak from Connecticut and Florida west to Minnesota and Texas.)

REFERENCES

Carvalho, J. C. M. 1952. On the major classification of the Miridae (Hemiptera) (with keys to subfamilies and tribes and a catalogue of the world genera). Anais. Acad. Brasil. Cien. 24:(1):31-110.

Carvalho, J. C. M. 1955. Keys to the genera of Miridae of the world (Hemiptera). Bol. Mus. Para. Emilio Goeldi XI: fasc. II:5-151.

Carvalho, J. C. M. 1957-1960. A catalogue of the Miridae of the world. Arq. Mus. Nacional (Brazil):44:1-158; 45:1-216; 47:1-161; 48:1-384; 51:1-194.

Knight, H. H. 1941. The plant bugs, or Miridae, of Illinois. Bull. Ill. Nat. Hist. Surv. 22:(1):1-234.

Knight, H. H. 1968. Taxonomic review: Miridae of the Nevada test site and the western United States. Brigham Young U. Sci. Bull. (Biol. Ser.) 9:(3):1-282.

ISOMETOPIDAE
The Jumping Tree Bugs

This is a very small family consisting of less than a dozen species north of Mexico. All are small insects less than 3 mm long with enlarged hind legs for jumping, with a cuneus present in the fore wing and one or two cells in the membrane. Isometopids are very closely related to the Miridae and probably actually represent only a very distinct sub-family. They are distinguished from the latter principally by the presence of ocelli which are lacking in the great family of the true plant bugs. Little is known of their biology but they apparently live on the bark and dead limbs of trees. Some species are known to be predaceous.

1   Cuneus and 2 cells of membrane extending almost to apex of wing .... *Diphleps*

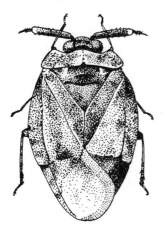

Figure 416

Figure 416 *Diphleps unica* Bergroth

This is a small mottled grayish-yellow and brown insect about 2 mm long with numerous dark spots scattered over the dorsal surface. The body is broad and flattened and resembles a tiny *Peritropis* (a mirid). The second antennal segment is very large and thick with black and yellow stripes.

This curious little insect has been taken on lichens and sometimes comes to lights. This species is apparently widely distributed although rare and known only from Connecticut, Maryland, Illinois, Ohio, Missouri and Iowa.

1a Cuneus and cell or cells in membrane never extending almost to end of wing 2

2   Eyes very large, occupying most of dorsal head surface, nearly meeting on midline in front of ocelli (fig. 417); head only about 1/2 width of hind margin of pronotum .... *Myiomma*

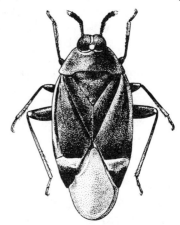

Figure 417

Figure 417 *Myiomma cixiiformis* (Uhler)

This is a remarkable little insect about 2.5 mm long with the head bent down on the ventral surface as in Homoptera, the resemblance to some of which the name implies. It is chiefly dull black with a pale head and a conspicuous white band across the base of each cuneus. Apparently it lives on bark and the twigs of dead trees. It is probably predaceous. It is known from New York south to Virginia and west to Michigan but it is rare.

2a Eyes relatively small, separated from each other by a distance at least half width of one eye; head more than 1/2 basal width of pronotum .... *Corticoris*

Corticoris pulchellus (Heidemann)

This is a small (2.2 mm) broadly oval species of a dark brown shining color with predominately white wings. The clavus is dark brown basally and has dark spots at the posterior end of each clavus and near the inner basal angle of the cuneus. It occurs in the eastern states from New York to Virginia.

Three additional species are known, one from the District of Columbia, the other two from the Southwest.

REFERENCE
Eyles, A. C. 1971. List of Isometopidae Heteroptera: Cimicoidea). New Zealand Jour. Sci. 14:(4):940-944.

DIPSOCORIDAE[11]

This is a small family of very minute hemipterans closely related to the Schizopteridae in many ways. They may be recognized by the presence of only two tarsal segments on all legs in the female, by the elongate second antennal segment which is at least twice as long as the first segment, by the asymmetrical male claspers and genital capsule, by the presence of numerous long bristles on the head and tibiae and by the porrect heads. They resemble schizopterids in having the fore coxae and the base of the head below enclosed by an enlargement of the prothorax (although different detailed structural elements are involved).

The dipsocorids live in ground litter and among stones. They are presumably predaceous. Unlike the Schizopteridae they do not jump when disturbed, but run actively over the surface of the ground. Only two genera are known to occur north of Mexico.

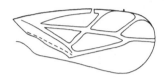

Figure 418

Ceratocombus vagans McAtee & Malloch

This is a tiny (0.75-1.75 mm) hemipteran of a pale to dark brown color with pale yellow legs and antennae. The front wings are brightly shining. These insects live among fallen leaves and ground litter. The known range is from New York to southern Florida.

A second species (*hesperius* M. & M.) is known from California and an additional species *C. latipennis* Uhler (fig. 419) possibly synonymous with *vagans* is known from New Mexico and Missouri.

1 Second antennal segment three or more times as long as segment one; "fracture" midway along costal margin of front wing not extending beyond the costal area. Fig. 418 .... *Ceratocombus*

11. There is a great deal of confusion in the literature concerning the correct name to use for this family. Many authors use the names Cryptostemmatidae and Ceratocombidae. However Dipsocoridae is the oldest higher group name and is the correct name to be used.

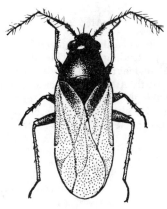

Figure 419

Figure 419 *Ceratocombus latipennis* Uhler

1a Second antennal segment less than three times as long as segment one; "fracture" midway along costal margin of fore wing extending far beyond costal vein to or nearly to center of wing. Fig. 420 .... *Cryptostemma*

Figure 420

Cryptostemma uhleri McAtee & Malloch

This species is dull yellow to brown in color, and from 1.25-1.5 mm in length. The body surface is shining and the second antennal segment twice the length of segment one. *C. uhleri*, like other members of the genus, lives under stones along stream margins. Adults are said to resemble small Collembola and to move very rapidly when disturbed. In this country *uhleri* has been taken only in the mountains of northern Georgia.

A second species (*C. usingeri* Wygodzinsky) is known from California where it was taken under rocks at the edge of hot pools.

SCHIZOPTERIDAE

This is an unusual family of extremely minute heteropterans that range in size from 0.8 to 2.0 mm. The front wings are convex, strongly sclerotized and beetle-like, although they overlap slightly along the midline. One of the most distinctive features is the development of the pleural region of the prothorax which is enlarged ventrally to enclose the fore coxae and also to enclose the ventral surface of the head posteriorly. The inner surface of the hind coxae has a pair of roughened pads which are used in conjunction with a metasternal spine as a jumping organ. In North American genera the males have three tarsal segments on each leg, whereas females have two segments on the front and middle legs and three-segmented hind tarsi. The male genitalia are bizarre and strongly asymmetrical. Antennal segments one and two are subequal in length.

The family is closely related to the Dipsocoridae but readily distinguishable by the long second antennal segment of the Dipsocoridae which is usually twice as long as the first segment. Nymphs of Schizopteridae have only one pair of dorsal abdominal scent glands whereas Dipsocoridae have three or four pairs.

These tiny insects most frequently occur in damp soil and in forest litter. They jump actively when disturbed. Little is known of their feeding habits, but they are thought to be predaceous.

Despite the fact that the family is primarily tropical and subtropical and very rare in the United States all three species thus far known from North America are included in an attempt to encourage collection of these remarkable insects by American students.

1. Labium four-segmented (but with segments three and four somewhat indistinct and probably incapable of independent movement); vein closest to trailing edge of wing curving forward and contacting vein immediately anterior. Fig. 421 .... *Glyptocombus*

Figure 421

Glyptocombus saltator Heidemann

This is a tiny black insect with the legs, labium and the basal segments of the antennae brownish yellow. The eyes are very large and extend backward over the antero-lateral portion of the pronotum. It is found on the ground among fallen leaves and litter.

This species is known in the literature only from Maryland and Tennessee, but specimens recently have been taken in Michigan (T. Schuh *in litt.*).

1a. Labium only three-segmented; vein closest to trailing edge of wing ending in membrane before apex, not curving laterad to meet next anterior vein. Figs. 424, 425 2

2. Labium truncate at apex. Fig. 422 .... *Corixidea*

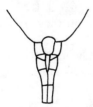

Figure 422

Corixidea major McAtee & Malloch

This species is easily recognized by the very large strongly protruding eyes. The labial segments are subequal in length, and the labium is distinctive among North American species of the family in being strikingly truncate at the apex. It is a blackish color with bluish-gray pubescence.

This species is known only from Tennessee.

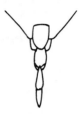

Figure 423

2a. Labium pointed at apex. Fig. 423 3

3. Second labial segment less than twice as long as segment three; vein separating first (basal) from second cell of fore wing along costal margin lying at right angle to wing length. Fig. 424 .... *Schizoptera*

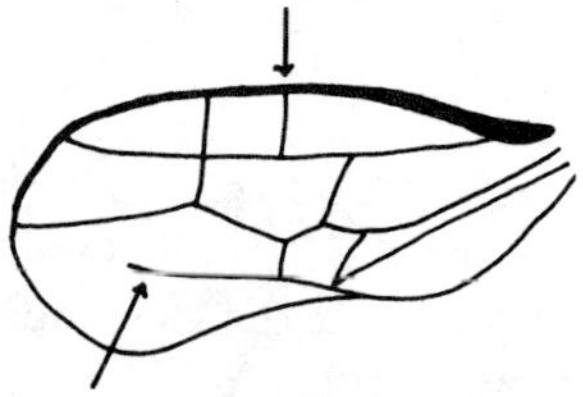

Figure 424

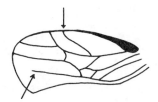

Figure 425

Schizoptera bispina McAtee & Malloch

This is an elongate oval species, nearly black in color, with the distal third of the fore wings a contrasting white. The legs are pale yellow.

In this country it is known only from Florida where it has been taken on Spanish moss, in litter and on sugar cane.

3a Second labial segment more than twice as long as segment three; vein separating first and second cells of fore wing along costal margin placed at an oblique angle to wing length. Fig. 425 .... *Nannocoris*

Nannocoris arenaria Blatchley

This species can be readily recognized by the elongate porrect head and by the very long second labial segment. The color of the body is dark brown with the legs, antennae and recurved costal margins of the hemelytra yellow. It apparently lives on the ground in a sandy habitat.

This species is known only from Florida.

REFERENCE
Emsley, M. G. 1969. The Schizopteridae (Hemiptera: Heteroptera) with the description of new species from Trinidad. Mem. Amer. Ent. Soc. N. 25:1-154.

HYDROMETRIDAE
The Marsh Treaders

This is a small family of very elongate slender stick-like insects with long legs and antennae that live on the surface of quiet water. Many students when first encountering these remarkable little insects believe they are tiny aquatic "walking sticks" which indeed they are in appearance although not in relationship. Marsh treaders can be recognized readily by their elongate bodies and thread-like legs, with the front legs slender like the middle and hind legs, and not adapted for grasping prey. The head is very elongate and the eyes appear as a pair of tiny round beads located midway from the tip to the base of the head. The claws are located at the tips of the tarsi and the wings are often reduced or absent even in the adults. They are usually found on the surface of small ponds with abundant vegetation where they walk sedately over the surface searching for small insects and ostracods upon which they prey.

In America north of Mexico only seven species are known and all belong in the genus *Hydrometra*. The commonest and most widespread species is *Hydrometra martini* Kirkaldy (fig. 426) which occurs throughout eastern North America and is found irregularly in the West. The other species all have a southern and western distribution.

Figure 426

REFERENCES

Torre-Bueno, J. R. de la. 1926. The family Hydrometridae in the Western Hemisphere. Ent. Amer. 7:83-128.

Drake, C. J. and D. R. Lauck, 1959. Descriptions, synonymy, and checklist of American Hydrometridae (Hemiptera: Heteroptera). The Great Basin Naturalist 19(2, 3):43-52.

GERRIDAE
The Water Striders

These are the most common and best known of the Hemiptera that live on the surface of the water. Many are known as "water spiders," "pond skaters," "wherrymen" or "good luck bugs" and are familiar to almost everyone. They dart actively over the water surface seeking small insects upon which they feed. This is a rather large family characterized by having the claws arising before the end of the tarsus, the body usually covered in large part with a thick coat of waterproof velvety hair, two-segmented tarsi and no scutellum. A single scent gland opening called an omphalium is located in the middle of the metasternum. Both wingless and winged forms are common for many species but some have only winged forms. Water striders occur on both still water and the swift parts of streams. Members of the extralimital (chiefly tropical and subtropical oceans) genus *Halobates* are the only truly pelagic insects.

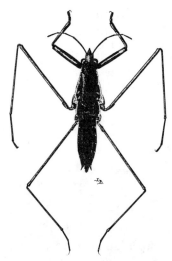

Figure 427

The members of the genus *Gerris* are the most common and widespread of the water striders in the United States. Nearly twenty species occur north of Mexico and many of these are widespread in the northern states. Identification is based in large part upon the nature of the margin of the last abdominal segment before the genitalia, and upon the male genitalia themselves. The following key is based only upon males and includes the most widely distributed species.

1   Inner margin of eyes conspicuously concave or sinuate behind middle; abdomen elongate 2

1a Inner margin of eyes evenly rounded throughout; abdomen usually very short 3

2   Dorsal surface of pronotum dull, thickly covered with short velvety hairs; first segment of fore tarsus only slightly shorter than second. Fig. 427 .... *Gerris*

Figure 428

This is a large (15-17 mm) species, brownish black in color, with extremely long legs. It is distributed from the northeastern United States south to North Carolina and west to Ohio and Michigan, and is common in the southern part of its range.

cc Last abdominal sternum before genital segments with a broad, deep median furrow. Fig. 430 ...... *G. nebularis* D. & H.

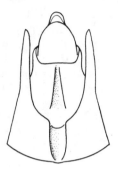

Figure 430

a Last abdominal sternum before genital segments evenly concave below (fig. 428), lacking a second concavity near middle (although in *nebularis* there is a deep median groove) .... b

aa Last abdominal sternum before genital segments with a second concave impression in central area. Fig. 429 .... e

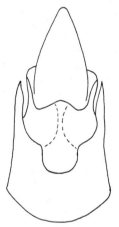

Figure 429

This is a long-legged species very similar to *conformis* but can be separated from it by a deep median groove in the last abdominal segment below (this makes the margin look superficially like those species with a double concave impression). It occurs throughout the eastern United States from New York west to Nebraska and south to Florida and Louisiana.

d Smaller, usually under 10 mm in length (8-11 mm) .... *G. canaliculatus* Say

This is a medium-sized relatively very slender, delicate species, brownish-black or reddish-brown in color. The connexival spines are very long. It occurs from New York west to Iowa and south to Florida and Louisiana but is common only in the south.

dd Larger, 12-15 mm in length .... *G. dissortis* D. & H.

b First antennal segment at least as long as second and third segments combined c

bb First antennal segment shorter than second and third segments combined ...... d

c Last abdominal sternum before genital segments lacking a conspicuous median furrow .... *G. conformis* (Uhler)

This is a large, predominately bright reddish-brown species with the anterior pronotal lobe having a large oval black spot on either side of the midline. It is a northern insect known from New England west to Iowa and south to Maryland and Missouri.

Families of Hemiptera

G. notabilis D. & H. is a related western counterpart of *dissortis*. It closely resembles *dissortis* but has longer legs and antennae, and the posterior margin of the first genital segment is "squared off" instead of being somewhat rounded. The distribution is from Iowa west to the Pacific coast.

    e   Size large, usually well over 11 mm. Fig. 427 .... *G. remigis* Say

This is by far the most abundant of our larger water striders. The body is heavier than most of the other large species and usually is a dark brown to black color. The anterior lobe of the pronotum usually has a narrow yellowish or orange stripe down the midline. It is found on ponds and lakes in the west and on rivers and streams in the east.

    ee   Smaller less robust species, size always less than 11 mm .... f

    f   Underside of first genital segment with numerous long silvery hairs in a tuft on each side of midline. Fig. 431 .... *G. comatus* D. & H.

Figure 431

This is a small dark brown or black species without light markings. Males are easily recognizable by the tufts of long hair on the first genital segment. The length is from 7-9 mm. It is widely distributed from New England west to Montana and south to Virginia, Kansas and Colorado.

The closely related *G. incognitus* D. & H. is common on the West Coast. There is, however, a record from Quebec. It has small genital hair tufts that originate on the midline and has a pale brown stripe anteriorly along the lateral pronotal margin.

    ff   First genital segment lacking tufts of long silvery hairs .... g

    g   Lateral margins of pronotum anteriorly with a silvery or reddish-brown stripe .... *G. buenoi* Kirkaldy

This is a short stocky water strider (7-8.5 mm) with brownish or yellowish stripes anteriorly on the lateral margins of the pronotum and also on the median line of the anterior lobe. It ranges across the continent in southern Canada and the northern states.

G. argenticollis Parshley is closely related and seems to replace *buenoi* in the central states and southward from Massachusetts and Illinois to North Carolina and Louisiana. It is difficult to distinguish from *buenoi* but has the first genital segment distinctly longer than broad whereas in *buenoi* the first segment is about as broad as long. It is always macropterous.

    gg   Antero-lateral pronotal margins lacking a pale stripe .... *G. marginatus* Say

This is frequently the most common of the smaller members of the genus and is found throughout the United States. The appearance of the first genital segment which is strongly impressed on each side will separate *marginatus* from the very similar appearing *G. insperatus* D. & H. which also is widespread over the eastern United States and west to Colorado and Texas.

The status of some of these closely related species seems to us to be in need of careful study. A number of species seem to replace each other geographically and an analysis of large samples should bring rewarding information as to whether or not we are dealing with infraspecific populations.

    2a   Pronotum with relatively small scattered hairs present; dorsal surface of pronotum shining; first segment of fore tarsus much shorter than second .... **Limnogonus**

Limnogonus hesione (Kirkaldy)

This is a predominately black shining species with a relatively shorter abdomen than in species of *Gerris*. It has a yellowish or brownish area at the base of the head and a large spot mesally on the anterior pronotal lobe. The posterior and lateral margins of the posterior pronotal lobe are also usually pale margined. This species is usually found in schools on lakes and ponds. The range is throughout the eastern United States west to Nebraska and Texas. An additional species occurs in the southwest.

3   First antennal segment very long, as long as or longer than the other three segments together. Fig. 432 .... *Metrobates*

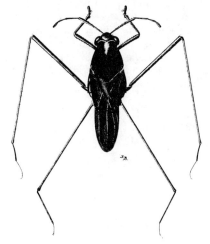

Figure 432

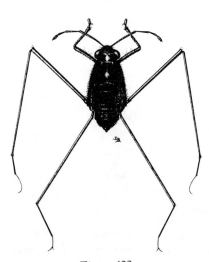

Figure 433

Figures 432 and 433   *Metrobates hesperius* Uhler

This is a small (3-5 mm) water strider with a very thick short body giving it an almost oval appearance. The pronotum is very small. The color is velvety black with the mesonotum marked with three grayish stripes that become very dull in dead specimens. The vertex of the head is bright orange. The coxae and trochanters are yellow and there are yellow markings at the base of the first antennal segment, between the eyes and in a median depression on the pronotum. Both macropters (fig. 432) and apters (fig. 433) occur. This species is usually found in the quiet parts of streams and rivers. It ranges over most of the eastern United States west to Kansas and south to South Carolina.

Five additional species occur in the southern and western states.

3a First antennal segment much shorter than the three remaining segments together 4

4   Third antennal segment with several stiff bristles present; antennae of males often much enlarged and curved (if antennae simple then body length not exceeding 2.25 mm) .... *Rheumatobates*

Figure 434

Figure 434   *Rheumatobates rileyi* Bergroth

The males of these tiny (2.5-3 mm) water striders are perhaps the most bizarre of any of our North American Heteroptera. Not only are the antennae much thickened and curved with the fourth segment bearing a pair of sharp spine-like hooks, but the hind femora are shortened, greatly thickened and twisted, and bear projecting processes. The middle femora bear a fringe of extremely elongate hairs. The coloration is velvety black with white or yellow median spots on the pronotum and often on the mesonotum. The entire mesosternum is yellow. They are most common in running water but have also been taken in lakes and ponds. The range is over the northeastern states west to Minnesota and south to North Carolina. When found they are sometimes abundant but seem to be rather local in occurrence.

Six additional species are known (*R. minutus* Hung., a tiny species from Florida, has the male antennae unmodified).

4a Third antennal segment bearing only fine hairs; antennae of both sexes simple and never greatly thickened and curved; body length always well over 2.23 mm .... *Trepobates*

Figure 435 *Trepobates pictus* (H. S.)

This species has a short rounded body form but is less ovoid than *Metrobates*. The body color is a mixture of black and yellow stripes, bars and spots and is extremely variable. The head is usually light with a longitudinal median black patch. The fore femora are usually yellow streaked with black while the tibiae and tarsi are dark. The length is from 3.5 to 5 mm. It ranges widely throughout the eastern states west to Illinois but is scarce in the extreme south. It is common on ponds. Adults superficially resemble nymphs of some species of *Gerris* and are thus often overlooked.

Eight additional species occur north of Mexico.

REFERENCE
Drake, C. J. & H. M. Harris, 1934. The Gerrinae of the Western Hemisphere. Ann. Carnegie Mus. 23:179-240.

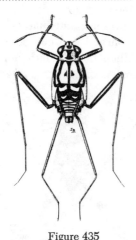

Figure 435

VELIIDAE
The Smaller Water Striders

This is a family of small to minute species characterized by having rather short bodies, the pronotum and mesonotum fused, and the scent gland openings with laterally extended channels. Wings are seldom present. These small insects live on the surface of the water or the adjacent shore where they move about actively seeking small insects upon which they feed. Members of the genus *Rhagovelia* are known as "riffle bugs" and are usually found in "schools" darting about in fast running streams where the current is strongest. Members of the genera *Paravelia* and *Microvelia*, on the other hand, are usually found in the quiet backwaters of streams and on ponds. There are several remarkable marine genera and some tropical species that live in bromeliads.

1 Middle tarsi deeply cleft with flat plate-like claws; often with plumose hairs arising from base of cleft. Fig. 436 .... *Rhagovelia*

Figure 436

Figure 437

Figure 437 *Rhagovelia obesa* Uhler

This species is black above with the basal 1/3 of the first antennal segment pale yellowish white. The pronotum has a small orange spot or patch on each side of the midline near the anterior margin. The male pronotum is rounded posteriorly and does not project backward to completely cover the mesonotum mesally. The posterior margin of the female pronotum is slightly concave and never produced into a projecting process. The abdomen of the female is very different from that of the male, having the lateral abdominal margins expanded and curved upward and over the abdomen to nearly meet at the midline. The middle femora are much longer than the hind femora. The length is 3-4 mm. It is widely distributed from southern Canada south to Alabama and perhaps Florida and west to Illinois and Minnesota.

R. oriander Parshley is similar in size and color. Males differ from those of *obesa* in having the posterior end of the pronotum conspicuously triangular and projecting back over the metanotum. Females of *oriander* have the pronotum projected backward in a blunt knob. The distribution is midwestern.

R. distincta Champion is larger than *obesa* (over 4 mm). Females have the connexiva distinctly separated along the midline (contiguous in *obesa*). Males differ from *obesa* in having the posterior margin of the mesonotum narrowly but distinctly exposed beyond the pronotum. This is a common species in the western states.

Eight additional species occur north of Mexico.

1a Middle tarsi not deeply cleft and lacking a plume of hair 2

2 Fore tarsi one-segmented; first antennal segment not longer than segment four .... *Microvelia*

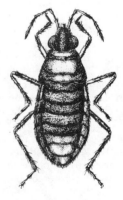

Figure 438

Figure 438 *Microvelia pulchella* Westwood

The members of this genus are the smallest of the water film living Hemiptera. This is a very small (1.5-2 mm) dark brown species that is velvety on the head and pronotum. It is marked with yellow at the base of the first antennal segment, on the front of the pronotum, the middle of the abdomen and the basal half of the coxae. Apterous forms are much the most common. This is an abundant insect on ponds covered with duckweed where these minute creatures often move in great numbers on the surface film. The distribution is almost throughout the United States.

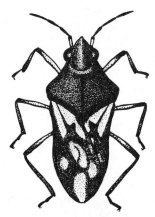

Figure 439

M. hinei Drake (fig. 439) is an even smaller species, only about 1.5 mm in length. In *M. hinei* the pronotum extends backward over the mesonotum to the region of a pair of lateral pits (*M. pulchella* has a short pronotum). The color is dark brown. This also is a wide-ranging species that is distributed over much of the United States.

M. americana (Uhler) is up to 2.5 mm long and has the first and third antennal segments equal in length and the mesonotum distinctly visible. The dorsal surface of the abdomen has conspicuous silvery patches of short hairs. The macropterous form has the wings nearly uniformly brown. This species occurs from the eastern states west to Colorado and Utah.

M. albonotata Champion is easily recognizable in the macropterous condition by having a conspicuous white stripe on each clavus and corium and an oval white spot on each corium. The membrane has three white spots. In the wingless form the abdomen posterior to the first two brown segments is pale greenish. The first antennal segment is shorter than the third. It is distributed throughout the United States east of the Rocky Mountains.

There are 13 additional species known from north of Mexico.

It is sometimes difficult to tell wingless adults of these tiny insects from the nymphs. However, nymphs have the tarsi of all legs one-segmented whereas adults have two-segmented middle and hind tarsi.

2a   **All tarsi three-segmented; first antennal segment much longer than any other segment** .... ***Paravelia***

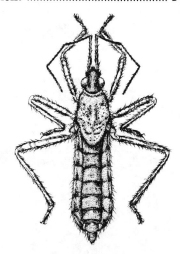

Figure 440

Figure 440   *Paravelia stagnalis* (Burmeister)

This is a bright brown or brownish yellow, nearly parallel-sided species. The underside of the body has a broad black stripe that runs from near the front legs to the sixth abdominal segment. The length is 4 mm. The middle and hind tarsi are equal in length with the second and third segments of the middle tarsi about the same length. It is found in the eastern United States north to Pennsylvania and Ohio but is usually uncommon. Our American *Paravelias* seem to often occur on the ground near still water, about grass roots and in masses of floating vegetation.

P. brachialis (Stål) is a more heavy-bodied insect, often over 5 mm in length, with the middle tarsi longer than the hind tarsi and with the middle tarsi having the second segment much longer than the third. The tibiae and femora are ringed with white or yellow bands. It is found in the extreme southern states from Florida to Texas.

Two additional species are known from the southwestern United States.

REFERENCE

China, W. E. and R. L. Usinger, 1949. Classification of the Veliidae (Hemiptera) with a new genus from South Africa. Ann. Mag. Nat. Hist. Ser. 12:2:343-354.

MESOVELIIDAE
The Water Treaders

This is a small family of semi-aquatic bugs that live on the surface of lakes and ponds, especially those having a heavy growth of surface vegetation. All of the species are small insects from 2-4 mm in length. The claws arise from the end of the tarsi. Ocelli are present in fully winged forms but absent in wingless individuals. All members of the family are presumably predaceous upon small insects.

1   Pronotum with posterior margin straight or evenly concave (fig. 441), never strongly produced backward; scutellum in macropters prominently visible in dorsal view .... *Mesovelia*

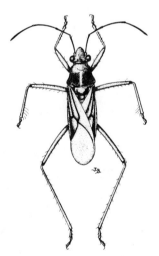

Figure 441

Figure 441   *Mesovelia mulsanti* White

This is the largest (3.5-4 mm) and by far the most common of our water treaders. It is most common in the apterous condition and green to yellowish-green in color. Macropters generally have the posterior pronotal lobe dark brown with a pale yellow median stripe. The hemelytra are largely white with the lateral third and the corial veins strongly contrasting brown. The white membrane often has a narrow distal median dark brown stripe. This is a very common species that occurs over almost the entire United States.

M. amoena Uhler is a smaller species, less than 3 mm in length. It lacks spines on the hind margin of the front and middle femora which are present in *mulsanti*. The pronotum is brown; the head yellow with two stripes present. This minute insect is more cryptic in its habits than is *mulsanti*. It is usually found at the margins of pools in crevices and in debris. The distribution is southern from Florida to California but a species from Michigan described as *douglasensis* is by some specialists considered to be synonymous. Three additional species apparently are represented north of Mexico.

1a   Posterior margin of pronotum prominently produced posteriorly to completely cover scutellum even in macropters. Figs. 442, 443 .... *Macrovelia*

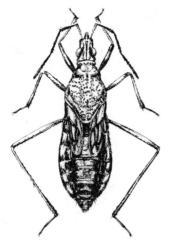

Figure 442

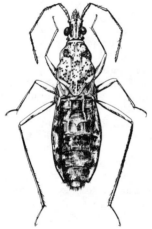

Figure 443

Figures 442 and 443 *Macrovelia horni* Uhler

This is a yellowish-brown insect about 4 mm long. The pronotum has a very distinct collar, has the anterior lobe swollen on either side of the midline and the posterior lobe covered with dark brown spots. Micropterous and brachypterous (fig. 442) forms appear to be the most abundant, the former having the wings reduced to small pads that extend only a short distance beyond the posterior margin of the pronotum (fig. 443). Macropters have six large cells in each hemelytron. The habitat is near permanent springs and streams where it is usually found in moss within a few feet of water and is at times taken on the water itself. The range is western from the Dakotas south to New Mexico and Arizona and west to California.

An additional genus and species (*Oravelia pege* D. & C.) which is apterous and lacks ocelli has recently been described from California.

HEBRIDAE
The Velvet Water Bugs

This is a very small family of minute insects that live on the water film of ponds or in damp soil near the water's edge. One finds them close to the shore in areas of abundant vegetation or often on shallow vertical overhanging banks but the extremely small size often causes them to be overlooked by the general collector. The hebrids are all presumably predaceous upon small organisms.

The family is characterized by the presence of four- or five-segmented antennae, a four-segmented beak, presence of ocelli, a small but definitely visible scutellum, and especially by the two-segmented tarsi with terminally placed claws. Hebrids may be either macropterous, brachypterous or apterous. When wings are present and fully developed the membrane lacks distinct veins.

1 Antennae four-segmented ...... *Merragata*

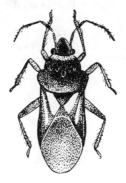

Figure 444

Figure 444   *Merragata hebroides* White

This is a tiny blackish insect (1.5-2 mm) marked with reddish brown on the head, pronotum and the antennae. The legs are yellowish. The front wings have the clavus and a large part of the corium white; the raised corial veins and the extreme corial apex are dark brown. The membrane is dark brown usually with 4 large white spots present. The pronotum has a number of pits on the anterior half and is very irregularly swollen and furrowed. *M. hebroides* is usually found on the surface of floating mats of aquatic plants. It is a very common species and is distributed in suitable habitats nearly throughout the country.

Two additional species occur north of Mexico.

1a Antennae five-segmented .... *Hebrus*

Figure 445

Figure 445   *Hebrus sobrinus* (Uhler)

This is a tiny (2-2.5 mm) reddish-brown species with yellow legs. The membrane has a pale yellow spot on either side near the base and usually three dull white spots near the center. This tiny water bug is usually found along stream margins in damp places rather than on the water surface itself. The distribution is over almost the entire country but the synonymy has been much confused in the past which makes the distributional data unreliable. It is largely replaced in the East by *H. burmeisteri* L. & S., *H. concinnus* (Uhler) and *H. consolidus* Uhl. However, separation of these species is really reliable only by the use of genitalia.

Approximately eleven additional closely related species occur in this country.

REFERENCE
Drake, C. J. and H. C. Chapman, 1958. New neotropical Hebridae, including a catalogue of the American species. Jour. Wash. Acad. Sci. 48(10):317-326.

LEPTOPODIDAE

This is a small family related to the Saldidae. The eyes are very large and protrudent and our species has a number of large spines projecting from the eyes. The third antennal segment is very elongate and slender. The labial segments, fore tibiae and femora as well as the head, pronotum and corium also bear prominent upright spines. The tarsi are three-segmented. This is an old world family which is represented in the Western Hemisphere by a single introduced species.

Figure 446

Figure 446 *Patapius spinosus* (Rossi)

This is a small (3.25 mm) yellowish-brown insect with a pair of dark brown transverse bands across the hemelytra. It is easily recognizable by the large spined eyes and labium. It lives on the ground, frequently distant from semi-aquatic habitats and is presumably predaceous on small insects. This is a south European species that was first reported in this country from California in 1941, and is now known to occur from Los Angeles County north to Butte County.

REFERENCE
Usinger, R. L. 1941. A remarkable immigrant Leptopodid in California. Bull. Brook. Ent. Soc. 36:164-165.

SALDIDAE
The Shore Bugs

This is a small family the members of which are generally recognizable by the black and white, or brown and white coloration, ovoid bodies and large protrudent eyes. The family has always been of much interest to students of the evolution of the Hemiptera as the species appear to show relationships to both terrestrial and aquatic groups. The antennae are four-segmented and prominent. The labium is elongate and only three-segmented. The membrane of the front wing is very distinctive in having either four or five elongate closed cells.

Saldids are most abundant near water on sand, mud flats and in salt marshes. However, some of the most interesting species occur on stones in rapid streams, in bogs and marshes and about springs. They are very active insects that generally fly readily and frequently and are a real challenge to the collector. Many resemble tiger beetles in their ability to pounce upon small insect prey or to escape the collector at the last possible moment. So far as known, all species are predaceous. The ecological requirements of many species appear to be very precise and careful ecological studies will unquestionably yield much important information.

1   Pronotum bearing two large thick upstanding tubercles on anterior lobe. Fig. 447 .... *Saldoida*

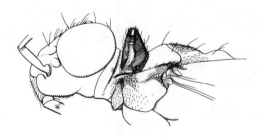

Figure 447

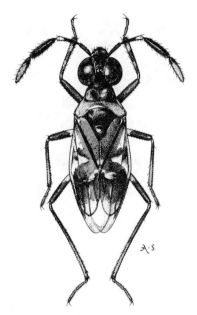

Figure 448

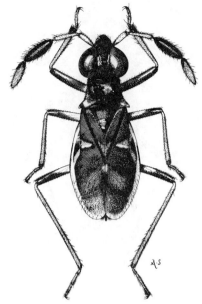

Figure 449

Figure 448 *Saldoida slossoni* Osborn

This is a small insect (3 mm) of a brown color and attractively marked with black, red, yellow and white. The species is an ant mimic and has been found associated with ants running about on the ground in moist places. The members of this genus are easily recognized by the ant mimicry, the white tipped antennae (very conspicuous in the field) and by the large pronotal tubercles. *S. slossoni* occurs from New Jersey south to Florida and westward along the Gulf Coast well into Texas.

S. cornuta Osborn (fig. 449) is a closely related Florida species distinguishable by the possession of acute backward projecting humeral pronotal angles (fig. 450) and non-elevated scutellar apex. *S. turbaria* Schuh has recently been described from sphagnum bogs in Michigan.

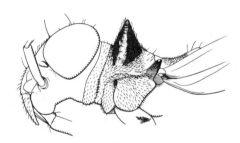

Figure 450

1a   Pronotum lacking a pair of large thick upward projecting tubercles on anterior lobe 2

2   Membrane of fore wing with 5 closed cells (fig. 451) (always fully winged) .... *Pentacora*

Families of Hemiptera   221

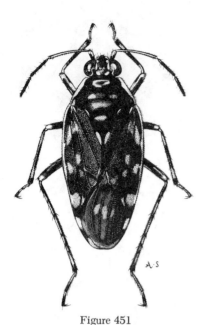

Figure 451

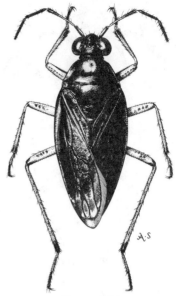

Figure 452

Figure 451 *Pentacora ligata* (Say)

This is a rather large saldid (5.5-6 mm) of a generally black coloration marked with a series of yellow spots scattered over the dorsal surface. The legs are yellowish with black bands. It is usually found on rocks in streams and is extremely agile. The range is over the eastern and central United States and at least as far west as Colorado and south to Georgia, South Carolina and Texas.

Four additional species occur in North America and are most common on beaches, but some also occur at inland localities.

2a Membrane of fore wing in fully winged forms with only 4 elongate closed cells (fig. 454); (wings sometimes reduced) 3

3 First or inner cell of wing membrane produced anteriorly (basally) 2/5 to 1/2 its length beyond base of second cell (fig. 452) (if cell somewhat less produced then antennal segments 3 & 4 strongly swollen) .... *Salda*

Figure 452 *Salda (Salda) provancheri*
Kelton & Lattin

This is a large shiny black shore bug frequently having a "coleopteroid" appearance. The antennae are generally dark with the inner surface of the first segment pale brown. The distal third of the fore wing bears 6 or 7 pale yellowish spots. It inhabits marshy areas, where it usually hides at the base of mosses or on clumps of sedges. The species is widely distributed through the eastern United States and west to Colorado and in the mountains to Arizona. Until recently it has been known by the name of *bouchervillei* (Provancher).

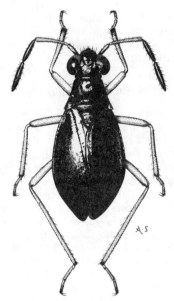

Figure 453

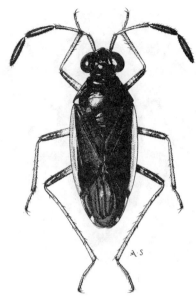

Figure 454

Figure 453 *Salda (Lampracanthia) crassicornis* (Uhler)

This is a small (4-5 mm) shining black species with strongly contrasting yellowish legs. It can be easily recognized by the strongly swollen third and fourth antennal segments. This striking shore bug is usually coleopteroid with the black shiny wing covers resembling the elytra of beetles. It sometimes occurs in damp meadows among tall sedges and grasses. Until recently it was considered to represent a distinct genus, but has recently been reduced to subgeneric status. The distribution is apparently the northern states and southern Canada, but it has been infrequently collected and the distribution is poorly understood.

Eleven additional species occur in North America.

3a Inner cell of wing membrane usually produced only slightly (never more than 1/3) forward of base of second cell. Fig. 454 4

4 Antennae relatively thick, third and fourth segments thicker than distal end of segment two. Fig. 454 .... *Ioscytus*

Figure 454 *Ioscytus politus* (Uhler)

This is a handsome species easily recognizable by the bright red corium and first and second antennal segments that contrast strikingly with the black clavus, pronotum and third and fourth antennal segments. It is widely distributed, and often common, in the western United States.

Four additional species occur in North America.

4a Antennae relatively slender, third and fourth segments not thicker than distal end of second antennal segment. Fig. 455 .... *Saldula*

Families of Hemiptera   223

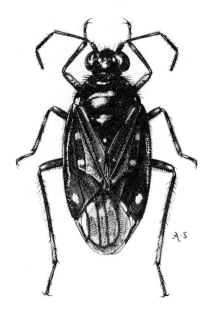

Figure 455

Saldula (Saldula) pallipes (Fab.)

This is our commonest shore bug over much of the United States. The color is variable but usually the head, pronotum, scutellum and basal fourth of the hemelytra are black with the remainder of the corium and the membrane pale whitish or yellowish. The outer margins of the pronotum are always entirely black. Forms occur with nearly completely black front wings. It is found chiefly along the shores of ponds, streams and lakes where it is often extremely common. There is some question as to the correct name to be applied to this variable species, and even whether or not only a single species is involved. This common species ranges over most of the United States.

Saldula (Micracanthia) humilis (Say)

This is one of our smallest saldids and usually occurs on sandy and muddy areas along streams and pond margins. It is frequently taken at lights. It is black in color and usually has a conspicuous small white spot present near the tip of the clavus. The lateral margin of the corium is pale white except for a dark area near the middle. The membrane is white with two small brown spots present within each cell. The range is very extensive from Ontario and New England south to Florida and west to Nevada, California and Arizona.

Throughout much of the west *humilis* is in large part replaced by the closely related *S. (M.) quadrimaculata* Champion.

These two, and several additional species, have generally been placed in the genus *Micracanthia*, but apparently in the future this will be reduced to subgeneric status.

Saldula is much the largest genus of North American shore bugs and contains at least 45 species. Many of the species are very similar in size and general appearance and their separation is extremely critical.

REFERENCE

Schuh, T. 1967. The shore bugs (Hemiptera: Saldidae) of the Great Lakes region. Contr. Amer. Ent. Instit. 2:(2):1-35.

NOTONECTIDAE
The Backswimmers

This is a moderate-sized but abundant family of aquatic insects easily recognized by the curious habit of swimming and resting with the ventral side up. The hind legs are elongate and fringed with long hairs and serve as oars to propel the backswimmer through the water. The front wings are always fully developed and formed to resemble the keel of a ship. All species are active predators and feed upon insects including mosquito larvae, small crustacea and even small fish. When handled they often bite vigorously, inflicting a painful "sting," and were called water bees by early German writers. Upon occasion they become serious nuisances in swimming pools.

Backswimmers are most abundant in ponds and lakes but also are sometimes common in slow moving streams. Some genera such as *Buenoa* are usually taken in open water where they lie below the surface in equilibrium with the surrounding medium. Males of many of these species stridulate actively during courtship. Most species overwinter as adults and can sometimes be seen swimming actively below the ice.

To accurately identify backswimmers it is sometimes necessary to dissect the genitalia or remove the foreleg to see details of the stridulatory area and of the labium which lies enclosed between the legs. Three genera are known from America north of Mexico.

1 Claval commissure lacking a pit or excavated area near anterior end; antennae four-segmented 2

1a Antennae three-segmented; clavus with a pit or excavated area present at anterior end of claval commissure just beyond tip of scutellum. Fig. 456 .... *Buenoa*

The members of this genus are small backswimmers and are readily recognizable by the small size and narrow, delicately formed bodies that are somewhat conical in shape. They are usually found in ponds and are one of the few insect groups that have cells containing haemoglobin, which possibly functions in maintaining equilibrium with the surrounding water.

Fourteen species are found in the United States, the majority being restricted in distribution to the southern and southwestern states. Males of the more common and widely distributed species may be separated by the following key (females often can be identified only by association).

 a Width between eyes at posterior margin of head 1/2 or more than anterior width of head between eyes .... b

 aa Width of head between eyes at posterior head margin less than 1/2 width of head between eyes at anterior margin c

 b Lateral projections on third segment of labium longer than third labial segment .... *B. macrotibialis* Hgfd.

This is a rather small species (5.75-6.5 mm) with the pronotum distinctly tricarinate. The color is variable but usually is dull white with a black band along the basal 1/4 of the lateral margin of the front wing. This is a northern species known from Nova Scotia and Quebec west to South Dakota.

 bb Lateral projections on third labial segment equal in length or shorter than third segment of labium. Fig. 456 .... *B. margaritacea* Torre Bueno

Figure 456

This species is from 6-7.5 mm in length and is a dull white to dark brown color which varies considerably even within the same population. Some specimens will key to "c" where they may be distinguished as noted in the discussion of the following species. This is a common and widely distributed species found nearly throughout the United States with the possible exception of the extreme northern states and Florida.

 c Fore femora narrowed at apices, length of femur more than three times width at distal end .... *B. scimitra* Bare

This species is very similar to *margaritacea* in habitus. It averages somewhat smaller (5.5-6.5 mm) and is best distinguished by having a long scimitar-shaped series of stridulatory ridges (60) on the inner surface of the fore femur. In *margaritacea* the stridulatory area is small with only 15-18 ridges present. *B. scimitra* is widely distributed in the southern states from Florida to California and north to Kansas and Virginia.

cc Fore femora widened at apices, length of femur three times or less than three times distal width .... d

d Pronotum strongly inflated, its median length equal to width across humeral angles .... *B. limnocastoris* **Hgfd.**

This species resembles *macrotibialis* but has 6 large spines along the inner face of the fore tibia (3 in *macrotibialis*). The inflated pronotum is definitive. It ranges widely through the eastern United States from Quebec and Maine south to Florida and in the north west to Minnesota.

dd Pronotum not strongly inflated, its median length at most 3/4 humeral width. .... *B. confusa* **Truxal**

This is a small species (4.25-5.75 mm) rather similar to *macrotibialis*. It has been confused with several other species in the past, as indicated by the specific name. It is widely distributed east of the Mississippi River from Connecticut to Florida and west in the north to Alberta and Manitoba.

2 Eyes separated from each other along midline; middle femur with a pointed spine present before distal end .... *Notonecta*

This genus includes the largest and most common of the North America backswimmers. Nearly twenty species have been reported from America north of Mexico. The body is rather robust and the majority of species are predominately black and white, although black, yellow and beautiful orange-red species also occur. The more common and widespread species may be separated by the following key.

a Central keel or ridge on ventral surface of fourth visible abdominal sternum bare along central area, hairs present only on sides of keel. Fig. 457 .... b

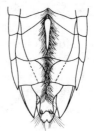

Figure 457

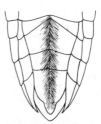

Figure 458

aa Central keel on under side of fourth visible abdominal segment covered with hairs even in central area. Fig. 458 c

b Minimum interocular space approximately 1/2 width of an eye; median plate (=metaxyphus) between anterior margin of hind coxae and posterior extension of middle coxae yellow; head length at least 1/2 length of pronotum. Fig. 459 .... *N. insulata* **Kirby**

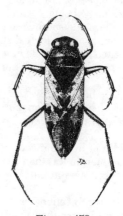

Figure 459

This is a large species (12.5-16 mm). The color is generally pale greenish yellow on the head and anterior half of the pronotum. The scutellum and usually the posterior half of the pronotum are black. The wings are pale grayish or dull whitish yellow with a black crossbar present beyond the apex of the claval commissure. Traces of reddish or orange color are often present, particularly distally on the corium. This species resembles a very large *undulata*. It seems to prefer cool water in pools. The range is over northeastern North America south to Maryland and west to the Great Plains.

    bb Interocular space usually much more than 1/2 width of an eye; plate between middle and hind coxae always at least in part black; head length less than 1/2 length of pronotum. Fig. 460 .... *N. kirbyi* **Hgfd.**

Figure 460

Figure 461

This is a common and easily recognized species by virtue of its large size and very dark coloration. It seems to prefer shaded pools and the backwaters of streams. *N. irrorata* is one of the very few North American species that inserts its eggs in plant tissues. It ranges over the entire eastern United States and west in the south to Arizona.

    cc Smaller species, usually not exceeding 11 mm in length; hemelytral coloration usually pale whitish, yellowish or orange, never with fore wings predominately blackish and mottled .... d

    d   Front wings orange-red and black; maximum width between eyes six times minimum width. Fig. 462 .... *N. uhleri* **Kirkaldy**

This species closely resembles *insulata* but is usually darker in color and is extensively marked on the fore wings with brown or dull brick-red coloration. It replaces *insulata* in the western states, being widely distributed from the Dakotas and Texas west to the Pacific.

    c   Large backswimmers, over 12 mm in length; front wings appearing dark throughout, extensively mottled with brown and yellow spots. Fig. 461 .... *N. irrorata* **Uhler**

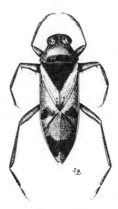

Figure 462

This is a very beautiful small species. The brick-red color of the hemelytra contrasts strikingly with the black scutellum and hemelytral crossbar adjacent to the tip of the claval commissure. The eyes are strongly convergent toward the base of the head and nearly come into contact with one another (fig. 462). *Uhleri* usually occurs in shady pools. It is often scarce and local in occurrence. The distribution is throughout the United States east of the Mississippi River.

> dd Front wings in large part white or yellowish, never deep orange; maximum width between eyes much less than six times minimum width .... e
>
> e Dorsal coloration except for eyes a nearly uniform white or whitish yellow; scutellum always chiefly pale. Fig. 463 .... *N. lunata* **Hgfd.**

Figure 465

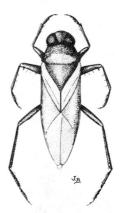

Figure 466

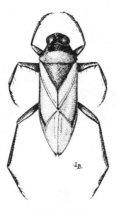

Figure 463

This is a small pale backswimmer that is rarely over 10 mm. The interocular space is narrow, usually only 1/3 the eye width. Occasional specimens have a dark crossbar on the wings, but they can be distinguished from *undulata* by the pale scutellum and distinctly angulate middle trochanters (fig. 464). This species is widely distributed east of the Mississippi River.

N. petrunkevitchi Hutchinson (fig. 466), known from the northeastern seaboard states, is very similar to *lunata* in size and coloration and the two sometimes occur together. *Petrunkevitchi* has a rounded angle to the mesotrochanter (fig. 465), whereas in *lunata* the mesotrochanter is distinctly angulate.

N. undulata Say (fig. 468), which like *petrunkevitchi* has a rounded mesotrochanter, sometimes has completely pale fore wings but is a more robust species and the scutellum is always in part black.

N. indica L. (fig. 467) also sometimes occurs in a completely pale form and also has a rounded mesotrochanter and then can be separated from *petrunkevitchi* only by the more robust size and a different shape to the genital capsule. The ranges of the two species do not overlap so far as is known.

> ee A dark crossbar present across fore wings beyond tip of claval commissure; scutellum in large part black .... f
>
> f Scutellum black with yellowish lateral and posterior margins; middle trochanters angulate .... *N. unifasciata* **Guerin**

Figure 464

This species resembles *undulata* in size and color but is readily distinguishable by the angulate mesotrochanters. It was first described by a French entomologist from specimens taken from a mixture of dried insects offered for sale in a Mexican market. It is a western species occurring on the Pacific coast.

ff Scutellum usually entirely black; middle trochanter rounded .... g

g   Median length of head (with specimen horizontal) greater than maximum interocular space. Fig. 467 ....
.... *N. indica* L.

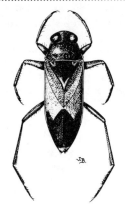

Figure 467

Figure 468

gg Length of head subequal to maximum interocular distance. Fig. 468 ....
.... *N. undulata* Say

This is our most common and most widely distributed backswimmer. It occurs in abundance in ponds and streams of many types. Its black and white color is variable but the scutellum is nearly always dark. It ranges throughout the United States but becomes scarce and local in the southern states and west of the Rocky Mountains.

2a Eyes broadly in contact with one another along midline; middle femur lacking a pointed protuberance or spine before apex .... *Martarega*

Martarega mexicana Truxal

This is our most common and most widely distributed backswimmer. It occurs in abundance in ponds and streams of many types. Its black and white color is variable but the scutellum is nearly always dark. It ranges throughout the United States but becomes scarce and local in the southern states and west of the Rocky Mountains.

This species closely resembles *undulata* in size and color but often has the dark wing markings more extensively developed, frequently as a complete black transverse band which will separate *indica* from "typically" colored specimens of *undulata* with which it is often confused. The distribution is across the entire southern United States usually below 37° N. latitude. (See *lunata* discussion for pale white individuals.)

REFERENCES

Hungerford, H. B. 1933. The genus Notonecta of the World. U. Kansas Sci. Bull. 21:(1): 5-195.

Truxal, F. S. 1953. A revision of the genus Buenoa (Hemiptera: Notonectidae). U. Kansas Sci. Bull. 35:(11):1351-1523.

PLEIDAE
The Pigmy Backswimmers

This is a small family of very tiny predaceous aquatic bugs whose members are usually less than 2.5 mm in length. Pleids are closely related to backswimmers (Notonectidae) and are included in that family by some authors. The species are yellowish or grayish in color, very convex and stout, with the wings forming a hard shell over the back. The labium is 3-segmented. In addition to the small size they can be distinguished from the Notonectidae by lacking a fringe of hairs on the hind tibiae and tarsi, by not having a keeled midline on the underside of the abdomen, and by having all three pairs of legs of nearly equal length.

Our only widespread common species is *Neoplea striola* Fieber (fig. 469) which is often abundant in ponds with a heavy growth of aquatic plants. It ranges over much of the eastern and central states and westward at least to Texas. Four additional species occur in the southeastern states.

There is a very closely related genus *Paraplea* two species of which occur in the southeastern states. *Paraplea* may be distinguished from *Neoplea* by having the front tarsi two rather than three segmented.

REFERENCE
Drake, C. J. & H. C. Chapman. 1953. Preliminary report on the Pleidae (Hemiptera) of the Americas. Proc. Biol. Soc. Wash. 66:53-59.

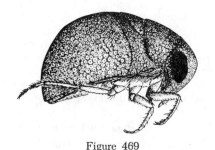

Figure 469

CORIXIDAE
The Water Boatmen

This is the largest family of the true water bugs. It not only contains the most species, but individuals often are extremely abundant. In this country they occur in nearly all types of water, even that with a high salt content. However, only a few species are found in running water. Water boatmen are most characteristic of ponds and lake margins. The family is easily recognized by the broad head with large transverse eyes that overlap the front of the prothorax; the highly modified "beak" that is broad, conical and unsegmented (a completely unique condition within the order); absence of ocelli; a fore wing that is of similar texture throughout and without veins in the membrane; scoop-like front legs; long slender middle legs with a single tarsal segment; hind legs flattened and fringed with long hairs for swimming. The water boatmen are indeed so greatly modified from other hemipterans in the nature of the beak and wings that some workers have considered them a distinct order or suborder. They apparently are able to take in small solid particles of food rather than liquids alone, and they feed on small organisms, both plant and animal (such as algae and even mosquito larvae) found in the water. They are thus the only known hemipterans able to ingest other than liquid material.

Sometimes water boatmen come to light in great numbers. These small water bugs are so abundant that they form an important part of the complex food chain of many ponds and lakes.

The accurate identification of water boatmen is often a laborious and difficult process. Many species and genera are distinguishable by the number and arrangement of the pegs on the scoop-like front legs and by the genital characteristics. The beginning student should not expect to determine his species without considerable effort and experience. The following key can in places be used only for males. Often study of the male genitalia is the only method for accurate identification. Eighteen genera and approximately 120 species occur north of Mexico. Many are northern in distribution and thus abundant in Canada and the northern United States. Frequently several related species will be found together in seemingly identical situations.

1   Labium lacking transverse grooves .... *Cymatia*

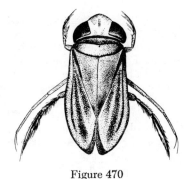

Figure 470

Figure 470 *Cymatia americana* Hussey

This is a slender species (5.8-8 mm) of a dark olive-green color fading to yellowish brown. The corium is spotted with numerous irregular black markings rather than bars or stripes. The pronotum has a medium longitudinal ridge on the anterior half. The pala of the front leg is long and slender and not scoop-like. It is flightless, the hind wings being reduced and non-functional. The distribution is northern from Michigan to British Columbia and northward to Alaska.

1a   Labium with a series of transverse grooves. Fig. 471 2

Figure 471

2   Fore tarsus slender and finger-like (fig. 472), tapering, with a large well developed apical claw present and projecting outward, much thicker than spines along lower margin of palm .... *Graptocorixa*

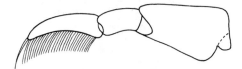

Figure 472

Graptocorixa californica Hgfd.

This is a rather robust species (8-10.5 mm). The dorsal surface appears roughened. The pronotum has about 18 alternating pale and dark transverse stripes of about equal thickness. The pale cross bands on the wing membrane have a wavy appearance. Both sexes have a large patch of long silvery hairs on the face. It occurs in running streams. It is widely distributed in California and Oregon.

Five additional species occur in the western states.

Figure 473

2a Fore tarsi often broad and spatulate (fig. 473), if pointed at distal end then with apical spine not conspicuously thicker than preceding spines along lower margin of pala, or if thicker then projecting downward rather than outward 3

3   Hemelytra lacking a distinct pattern; males with vertex of head pointed and with a prominent raised median ridge; male pala with upper surface bearing a broad deep notch. Fig. 473 ....
.... *Ramphocorixa*

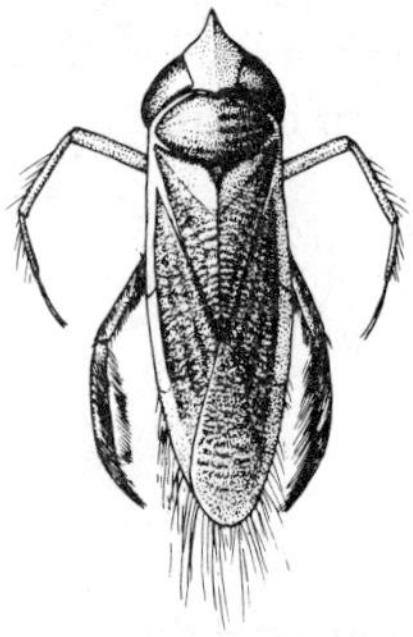

Figure 474

Figure 474 *Ramphocorixa acuminata* (Uhler)

This is an easily recognized species by virtue of the obsolete (or absent) color pattern which gives the species a nearly uniform pale gray or brownish coloration. The head and legs are whitish or yellow. The head of the males is acutely pointed with a prominent longitudinal carina, but females have a blunt and rounded head. The fore wings have the appearance of being semitransparent as compared with most other water boatmen. The length is 5-5.5 mm.

The most remarkable feature of this little species is its habit of laying eggs on the bodies of living crayfish that are associated with the boatmen in water holes. In the east it is not known north of Washington, D. C. and Ohio. Westward it occurs north to Minnesota and the Dakotas and extends to Colorado and New Mexico.

One additional species occurs in Arizona.

3a Hemelytra with a distinct pattern over at least a considerable portion of surface; males without anterior end of head pointed and vertex lacking a distinct elevated ridge; male pala not deeply notched along upper surface 4

4   Small shining species less than 5.5 mm in length; terminal portion of male abdomen asymmetrically modified to left. Fig. 475 .... *Trichocorixa*

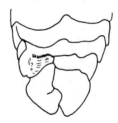

Figure 475

Trichocorixa calva (Say)

This species may be recognized by having the length of the polished area along the costal margin of the wing in front of the nodal furrow greater than the length of the middle tibiae. The interocular space is less than the length of the hind margin of the eye. The dark markings on the front wings form a distinct network. The pronotum has 8-9 dark transverse bars. This is a fresh water species and is frequently very common. It is distributed throughout the eastern and central states west to the Dakotas, Kansas and Arizona.

Trichocorixa verticalis (Fieb.) is a small shining species with the clavus and corium smooth and polished. The vertex of the head is rounded and prominently produced in front of the eyes. The outer margin of the fore wing is abruptly expanded somewhat before the middle. The fore wings have scattered long hairs and short spiny setae present.

The interocular space usually does not exceed the width of an eye measured along the hind margin. The coloration is very variable and five subspecies are recognized.

Verticalis shows a preference for salt and brackish water and the range is unusual. It is found along the entire east and gulf coasts, in Quebec near the mouth of the St. Lawrence, in southern California around the San Francisco Bay area and in interior North America through the Great Plains north to Hudson's Bay, but it is absent in the non-coastal East and in the Midwest, and also in the West except as noted above. The interior distribution is in salt flats and alkaline ponds.

T. louisianae Jacz. has a noticeable tuft of hairs at the apex of each clavus. The pronotum has 7-9 dark transverse bars that are much wider than the intervening yellow ones. The strigil on the dorsum of the abdomen is noticeably widened laterally (fig. 476). It is said to prefer salt or brackish water and is distributed along the Atlantic and Gulf Coasts from Massachusetts through Texas into Mexico and the West Indies.

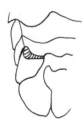

Figure 476

T. kansa Sailer is closely related to *louisianae* with the females being very similar, but *kansa* lacks the tuft of hairs near the apex of the clavus. In males the peg row on the pala is arranged in the form of an inverted "V" (fig. 477) and the anterior and posterior margins of the strigil are parallel to one another. The dark transverse pronotal bars are narrower than the light ones. The range is predominently midwestern, west to the 100th meridian and east to Pennsylvania and Florida.

Figure 477

T. sexcincta (Champion) (=*naias* Kirk.) has 6-7 dark transverse bars on the pronotum that are about twice the width of the pale stripes. The palar pegs are curved near the apex and obliquely arranged (fig. 478). It ranges widely over the entire eastern United States west to Colorado and Texas.

Figure 478

T. macroceps Kirk. is a very small species (2.8-3.2 mm) and is readily distinguishable because of its very short pronotum which is only one-fourth the pronotal width. *Macroceps* also lacks a nodal furrow on the embolar area of the wing. It is flightless and has only tiny vestiges of the hind wing present. The distribution is throughout the eastern United States west to about the 100th meridian.

Five additional species have been reported from America north of Mexico.

4a Usually larger than 5.5 mm; either strongly shining throughout or not; terminal portion of male abdomen asymmetrically modified to right 5

5   Pruinose area at base of claval suture (a) short and broadly rounded distally, at most two-thirds as long as postnodal pruinose area (b). Fig. 479 .... **Hesperocorixa**

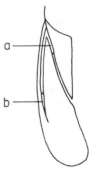

Figure 479

This is a genus of rather large species, many of them 8 and 9 mm in length. The pronotum and the fore wings usually show some rugose areas, and

Families of Hemiptera   233

the male front tibiae always has a bundle of spine hairs near the apex. There are 18 North American species and many of them are common and widespread.

Identification is not an easy task, but the more common species may be separated by the following key.

a   Mesepimeron (b) at level of scent gland opening (c) nearly as broad or broader than lateral pronotal lobe (a). Fig. 480 .... b

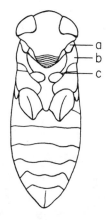

Figure 480

aa Mesepimeron at level of scent gland opening definitely narrower than lateral pronotal lobe .... e

b   Dorsal surface of hind femora with only 2-3 stout spines ....
.... *H. minorella* Hgfd.

This is a rather pale small (6-7 mm) species for the genus. The lateral prothoracic lobe is longer and more slender than in other species. The range is across southern Canada and the northern United States from Newfoundland and New England west to Minnesota and Manitoba.

bb Dorsal surface of hind femora armed with numerous spines c

c   Mesepimeron at level of scent gland opening about equal in width to lateral lobe of prothorax. Fig. 481 ....
.... *H. atopodonta* Hgfd.

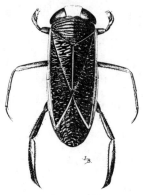

Figure 481

This is a large chestnut-colored species (8-9.5 mm). The pronotum has 9 broad straight dark stripes. Males have the last peg on the palar row set away from the others in contrast to all other species (fig. 482). There is a conspicuous yellow band along the apical margin of the corium. This is a very common species that ranges from Nova Scotia and New England south to New Jersey and west to Colorado and Washington.

Figure 482

cc Mesepimeron at level of scent gland opening much broader than lateral lobe of prothorax .... d

d   Less than 8 mm in length; corial pattern very dark, with cross stripes anastomosing to form a longitudinal series of stripes .... *H. semilucida* Hgfd.

In this species the color is dark and often a reddish hue is present. The pronotum has 7-9 dark cross stripes. The legs are frequently red; the underside of the abdomen is black with yellow or orange lateral markings. The distribution is eastern from Ontario and New England south to Florida and west to Illinois and Louisiana.

dd Length 8-9 mm, clavus and corium either solid black or with pale cross bands .... *H. kennicottii* (Uhler)

This is a large dark species with rugulose front wings. There is a tendency for this species to lose its light color markings and become entirely black. The range is from New England south to Virginia and west to Minnesota and Illinois.

    e   Length 7-7.5 mm .... *H. minor* **Abbott**

This is one of the smaller dark colored *Hesperocorixa*. The pronotum has 8 regular dark stripes tending to run together near the midline. The membrane of the fore wing has a light color pattern near its base but becomes "smoky" toward the tip. The head, legs and underside are yellow. This small species can be distinguished by the narrow mesoepimeron. *Minor* is distributed along the east coast from Massachusetts to Georgia and west to Texas in the Gulf states.

    ee   Length more than 8 mm .... f

    f   Color pattern of corium near lateral and posterior margins faded out and indistinct .... *H. lucida* (**Abbott**)

This is a large species (8.5-9.5 mm), usually without or with very faint stripes on the pronotum. The suture between the clavus and corium is marked with yellow. The incomplete corial markings are diagnostic. The distribution is throughout the eastern states west to Arkansas and Illinois.

    ff   Corium with color markings complete over entire surface .... g

    g   Pattern on front wings reticulate; pronotum smooth, without even faint rugosity .... *H. laevigata* (**Uhler**)

This is a dark colored large species (10-11 mm), with about 12 irregular stripes across the pronotum. This is the only *Hesperocorixa* with a pronounced reticulate pattern on the front wings. It is a very widely distributed species, occurring from southern New England to North Carolina in the east, and west to the Pacific. It appears to be most common in the far western states.

    gg   Pattern of front wings not reticulate; surface of pronotum rugulose, sometimes strongly so .... h

    h   Pale bands on corium posterior to hemelytral suture forming slender transverse series .... *H. vulgaris* (**Hgfd.**)

This is another large dark species (9-10 mm). The pronotum has about 10 dark stripes. The corium and membrane are clearly separated. The head and legs are pale and both fore wings and pronotum are strongly rugulose. This is a widely distributed species throughout the eastern United States west to the Dakotas and Texas but with scattered records from California and the northwestern states.

    hh   Pale bands on corium beyond hemelytral suture not forming slender transverse series, or if so then corium and membrane plainly separated i

    i   Hind femur bearing a row of about 10 spines below on outer portion of posterior margin .... *H. nitida* (**Fieber**)

The length is from 8-9 mm with the head, legs and ventral surface yellow. The pronotum has nine broad dark stripes. It is widely distributed in eastern North America from New England to northern Florida and west to Minnesota, Kansas and Texas.

    ii   Hind femur bearing a row of about 6 spines below on outer portion of posterior margin .... j

    j   Pattern of corium and membrane continuous, never separated by a pale stripe .... *H. obliqua* (**Hgfd.**)

This is a large species (10-11 mm) of a generally dark color. The pronotum has 12 narrow black stripes. The male pala is obliquely produced on the upper-outer angle. It is widely distributed in the eastern states and west to Colorado and Alaska, but the published records seem to indicate a rather spotty distributional picture that may be associated with strong habitat preferences.

    jj   Corium and membrane plainly separated, usually by a pale stripe .... k

    k   Interocular space nearly equal to width of an eye .... *H. lobata* (**Hgfd.**)

The pronotum in this large species (9.5-10.5 mm) is heavily rugulose and crossed with 8 dark stripes. The male fore tarsus has nearly parallel sides with a squared off apex and 28-30 teeth in the peg row. The distribution is from New England south to Georgia and across the northern states to Minnesota.

kk Interocular space considerably narrower than width of eye .... *H. interrupta* (Say)

This is another large dark species (9-11 mm) with 8-10 broad dark bands across the pronotum. The pronotum and fore wings are both finely rugose and the membrane shining. The male pala possess 28-30 teeth in the peg row. This is a common species ranging from eastern Canada south at least to Georgia and west to Nebraska and Arkansas.

There are six additional North American species of rather restricted distribution.

5a Pruinose area at base of claval suture attenuated, pointed or narrowly rounded at distal end, always more than 2/3 length of postnodal pruinose area 6

6 Posterior margin of head strongly curved posteriorly to contact widest portion of pronotum which laterally narrows to an acute, angulate point (fig. 483); interocular space much narrower than width of eye .... *Palmacorixa*

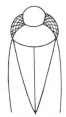

Figure 483

Palmacorixa buenoi Abbot

This is a small (4.5-6 mm) parallel-sided slender insect. The dark markings on the fore wings are very irregular and run together forming a reticulate or network pattern. The head is usually yellow. The pronotum has from 4-9 dark transverse stripes. The middle femur of the male lacks a longitudinal peg row below. This species is flightless with reduced wings. It usually occurs in permanent streams. The distribution is through the eastern and central United States west to the Dakotas and Texas.

Two additional species occur in the eastern and central states.

6a Posterior margin of head rarely reaching posteriorly to contact widest portion of pronotum, or if so then either interocular space not much narrower than width of eye or dorsal surface not rugulose 7

7 Dark markings on clavus occurring as complete straight non-interrupted bands 8

7a Dark markings on clavus interrupted, broken and often reticulate, never appearing as complete dark transverse bands 9

8 No male abdominal strigil present; male pala with 2 rows of pegs. Fig. 484 .... *Callicorixa*

Figure 484

Callicorixa alaskensis Hgfd.

This is a rather large water boatman (7-8 mm) and is of a general brown color. The pronotum has 9-10 narrow brown, transverse stripes. The first segment of the hind tarsus has the outer third black and strongly contrasting with the rest of the segment. A narrow pale brown band separates the corium from the membrane. The range is distinctly northern across Canada from Newfoundland to British Columbia. In the United States it is found only in the northern tier of states.

C. vulnerata (Uhler) is very similar in size, shape and color. The wing membrane is distinctly reticulate. The terminal part of the first tarsal segment of the hind leg is black as in *alaskensis*. Males of *alaskensis* have a dense patch of hair on the fore femur but in *vulnerata* the fore femur has only a scattered area of hairs. The distribution is in the northwestern United States and Canada south to central California.

C. audeni Hgfd. is distinguishable by the lack of a blackened third to the basal tarsal segment of the hind leg and by having the membrane of the front wing brownish yellow with an indistinct pattern. *Audeni* has a distribution similar to that of *alaskensis* vis: entirely across Canada and the northern tier of states.

Two additional species are found in North America, one in Texas and the other in the Aleutians.

8a   Male strigil present; male pala with one row of pegs. Fig. 485 .... *Sigara*

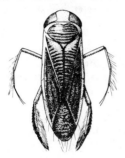

Figure 485

This is a very large complex genus of water boatmen containing nearly 50 North American species that have been subdivided into nearly a dozen subgenera. They vary from medium large to rather small species. The determination is complex and students wishing to pursue their determination should consult more advanced technical works.

Sigara alternata (Say) is the most common species. The length is from 5.5-7 mm. It is of a brown color, sometimes with a reddish tinge. The pronotum has 8-9 narrow brown bands. The coalescing of the dark coloration at the inner terminal angle of the corium is distinctive. It is abundant throughout the eastern United States west to the Great Plains and also occurs rarely in the far west.

9   Pronotum smooth and shining, at most with very fine obsolete longitudinal impressions; male pala triangular with 2 distinct rows of pegs .... *Corisella*

Corisella decolor (Uhler)

This is a rather small species (4-6 mm) with the pronotum bearing 6-8 dark bands much narrower than the pale ones. The markings on the fore wing are reticulate or network patterned, and the membrane suture is marked by an oblique pale stripe. The head is pale. The pronotum is very weakly roughened, and the interocular space very little greater than the width of the hind margin of an eye. It occurs in the western United States from the Pacific coast east to Colorado but with only a single record from the southwest.

C. edulis (Champion) is a larger species (6-8 mm) with the pronotum having 10-12 very narrow dark transverse stripes. Both the pronotum and fore wing are completely smooth and shining. The hemelytra bear short but distinct hairs. The range is from Arizona eastward across the southern states to Georgia and there are scattered records from Oregon, Minnesota, Iowa, Virginia and the District of Columbia.

C. tarsalis (Fieber) is a small species (6.5 mm) and readily recognizable by having both hind tarsal segments brown whereas in all of the other small species of *Corisella* at least the first segment of the hind tarsus is yellow. The general color is dull yellow with 10 narrow dark pronotal bars. The corium and membrane are reticulate. *Tarsalis* is a very widespread species from New York to California, but apparently is absent from the southeastern states.

Two additional species occur north of Mexico.

9a   Pronotum with distinct longitudinal ridges and grooves (rostrate); male pala broad with only one row of pegs .... *Cenocorixa*

Cenocorixa bifida (Hgfd.)

This is a medium-sized boatman (6.5-8 mm). The pronotum has a distinct median carina on the anterior third, and is rugose with about 10 irregular dark stripes. The head and legs are pale. The vertical surface is dusky to black. This is a northwestern species, most records being west of Colorado, but it has been collected in the northern tier of states as far east as Minnesota.

Eight additional species occur in the western United States.

REFERENCE

Hungerford, H. B. 1948. The Corixidae of the Western Hemisphere (Hemiptera). Univ. Kansas Sci. Bull. 32:1-827.

NEPIDAE
The Water Scorpions

This is a remarkable small family of aquatic bugs whose members are easily recognized by the elongate breathing tube that extends from the end of the abdomen like a slender tail. The forelegs resemble those of praying mantids in that the coxae are very much elongated and the femora and tibiae modified for seizing prey. The tarsi of all legs are single segmented. The middle and hind legs are slender and not flattened or fringed for swimming. Water scorpions spend most of their time clinging to sticks and weeds in ponds and streams waiting for passing insects or small fish to come within striking distance.

1   Body elongate and slender; hind coxae not separated from one another by a distance greater than width of a coxa .... *Ranatra*

Figure 486   *Ranatra fusca* P. B.

This large (35-42 mm) stick-like insect is dark reddish brown to dull yellow in color with a wide shallow groove in the prosternum. The legs lack or at most have very faintly darkened rings. The concealed antennae have a projection on the second segment. The front femora are constricted near the middle and have a distinct tooth near the tip. The pronotum is relatively stout, the front portion being less than twice as long as the enlarged basal portion. It is most common in weedy ponds. This is a common species found over most of the United States west to California and Texas.

R. nigra H. S., despite the name, is somewhat paler dull brown to yellowish and closely resembles *fusca* in size. It can be distinguished by lacking a tooth near the tip of the fore femur. The range is throughout the eastern United States west to Minnesota, Kansas and Texas.

R. kirkaldyi Torre Bueno is a smaller species (23-30 mm), usually with banded legs and always distinguishable by lacking a projection on the second antennal segment and in having fore femora that are not constricted near the middle and usually lacking a preapical tooth. The range is through the United States east of the Great Plains, but it is scarce in many localities.

Six additional species, mostly southern in distribution, are found north of Mexico.

1a   Body oval or elongate-oval; hind coxae separated from one another by distance much greater than width of a coxa 2

2   Body broadly oval, more than one-third as wide as long; propleura widely separated, leaving prosternum broadly exposed posteriorly .... *Nepa*

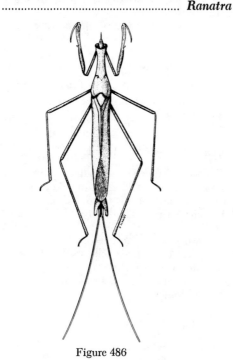

Figure 486

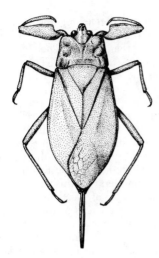

Figure 487

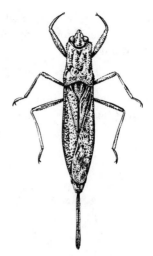

Figure 488

Figure 487 *Nepa apiculata* Uhler

This species is a dull grayish-brown color and somewhat resembles a belostomatid with a "tail" and non-swimming legs. The length is from 18-20 mm. It is found in ponds and sluggish streams either in the mud or in dense vegetation. *N. apiculata* is usually a scarce species but occurs over much of the United States east of the Great Plains. This is the only broad flattened water scorpion that occurs in much of the United States.

2a Body elongate-oval, less than one-fourth as wide as long; propleura contiguous over prosternum posteriorly ...... *Curicta*

Figure 488 *Curicta howardi* Mont.

This is a bizarre little water scorpion that looks superficially like a very much shortened, thickened *Ranatra*. The pronotum has a large deep pit on the posterior half near the lateral margin. The dorsal surface has a distinctly "warty" appearance. It occurs in shallow water in small streams and perhaps ponds. It is confined in distribution to the southwestern states. A second species is known from Arizona.

REFERENCE
Hungerford, H. B. 1923. The Nepidae in North America north of Mexico. Kans. Univ. Sci. Bull. 14:(1922):425-469.

BELOSTOMATIDAE
The Giant Water Bugs

This is a small family that contains some of the largest and most robust American insects. All species are very much flattened and elongate-oval in form and of a nearly uniformly dull brown to yellowish color. The front legs are greatly enlarged and raptorial. These insects are found in ponds, lakes and quiet waters of streams. All species are highly predaceous on aquatic insects and small fish. The hind legs are flattened, fringed with long hairs and are used as oars to move the bugs through the water. Some of the larger species come regularly to light, occasionally in very large numbers, and in some areas are known as "electric light bugs" or "toe biters." It is a credit to the interest these bugs arouse and an interesting sidelight on scientific objectivity to know that one small genus of giant water bugs has been carefully analyzed three times in recent years while scores of terrestrial Hemiptera genera of large size remain almost unknown.

1   First segment of labium very short, wider than long, and shorter than segment two .... *Lethocerus*

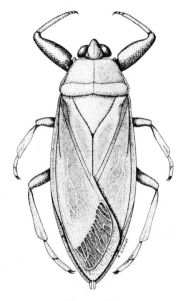

Figure 489

Figure 489   *Lethocerus americanus* (Leidy)

This is a very large brown insect (50-60 mm). The middle and hind femora have three, sometimes obscure, dark brown bands. The head is broad, with the minimum distance between the eyes usually equal to the maximum width of one eye. The fore femora have two deep grooves along the front margin and each fore tarsus has only a single claw. The range is nearly throughout the northern United States and southern Canada but it appears to be absent from the southeastern states.

L. uhleri is usually smaller in size than *americanus*, rarely exceeding 50 mm in length. The head is usually narrower, with the minimum space between the eyes not over three-fourths the maximum width of one eye, but one-fifth wider than the greatest width of the hind tarsi. The most reliable criterion for separation from *americanus*, however, is the adpressed hairs on the first ventral "laterotergite" not reaching the metapleuron and the disc of the mesosternum having minute spinules rather than hairs. The fore femora are grooved as in *americanus*. The distribution is throughout the eastern United States west to the Great Plains.

L. griseus (Say) is a large species (45-65 mm) readily distinguished from related species by lacking grooved fore femora. The legs lack definite dark and light stripes although they may be mottled. It is widely distributed through the eastern and central United States west to the Great Plains.

The members of this genus are the true "giant water bugs" of North America. Their eggs are not laid on the backs of the males, in contrast to the habits of other genera treated, but instead are laid in masses on emergent vegetation. Two additional species occur in the southwestern states.

1a First segment of labium much longer than wide and about equal in length to segment two 2

2   Membrane of fore wing well developed, its greatest width (with translucent margin) more than maximum width of clavus .... *Belostoma*

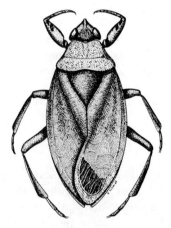

Figure 490

Figure 490   *Belostoma flumineum* Say

This is the most common of the smaller (22 mm) "giant" water bugs over much of the country. It is light brown in color with darker leg markings. The

240   Families of Hemiptera

side margins of the abdomen beneath (connexival plates) on segments two to five are completely covered with hairs. The head is rather short, the length before the eyes being equal to the length between the eyes. The range is over almost the entire United States, but it is relatively scarce in the southern and Pacific states.

B. lutarium (Stal) is a closely related species with the tylus produced forward more prominently, giving the head a more "snout-like" appearance, and it differs from *flumineum* in having the length of the head anterior to the eyes greater than the distance from the inner posterior angle of the eyes to the anterior edge of the eyes. There is a narrow area without hairs on the abdominal connexiva of segments two through five where these lateral plates meet the true ventral sternites. The range is over the entire eastern and central United States, and southwest to Texas and Arkansas. It is most common in the southeastern states, much more so than *flumineum*.

B. bakeri Mont. has a very short tylus, the length of the head in front of the eyes being only three-fourths the length of the head between the eyes. This is the most frequently collected species in the west and southwest.

Females of this genus have the surprising habit of gluing their eggs on the backs of the males.

Six additional species are known from the United States.

2a Membrane greatly reduced, usually to a small flap, usually not as wide as clavus .... *Abedus*

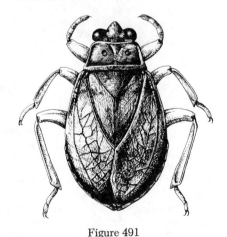

Figure 491

Figure 491 *Abedus indentatus* (Haldeman)

This species is longer than *Belostoma* species (27-37 mm), brown in color, with the wings very strongly oval, which gives the insect the appearance of tapering relatively strongly anteriorly. The abdominal segments below are completely covered with long silky pubescence. The antennae are three-segmented with short extensions from the second and third segments. It occurs chiefly in clear streams rather than ponds. The eggs are laid on the backs of the males. *A. indentatus* is found only in California where it is common.

A. immaculatus (Say) is the smallest belostomatid in the country (12-15 mm), and is a light brown color. The antennae are unique in being four-segmented with a projection from segment three. The end of the wing is more pointed than it is in other members of the genus. This is the only *Abedus* occurring in the eastern United States where it is a rare insect known from Florida, Georgia and Mississippi.

Three additional species occur in the southwestern states.

REFERENCES

Menke, A. S. 1963. A review of the genus Lethocerus in North and Central America, including the West Indies (Hemiptera: Belostomatidae). Ann. Ent. Soc. Amer. 56:261-267.

Lauck, D. R. & A. S. Menke 1961. The higher classification of the Belostomatidae (Hemiptera). Ann. Ent. Soc. Amer. 54:644-657.

Menke, A. S. 1960. A taxonomic study of the genus Abedus Stål (Hemiptera: Belostomatidae). Univ. Calif. Publ. Ent. 16(8): 285-288.

Cummings, C. 1933. The giant water bugs. Univ. Kans. Sci. Bull. 21:197-219.

Lauck, D. R. 1962-1964. A monograph of the genus Belostoma, Parts I-III. Bull. Chicago Acad. Sci. 11(3, 4, 5):34-154.

NAUCORIDAE
The Creeping Water Bugs

This is a small family of medium-sized oval dull yellowish brown colored water bugs that superficially resemble small belostomatids to which, however, they are not actually closely related. They are characterized by greatly enlarged raptorial front legs, the absence of ocelli, two segmented tarsi with long claws on the middle and hind legs and by the thick leathery fore wings that lack veins in the membrane. The legs are only weakly fringed with hairs and are not adapted for swimming.

These are rather inconspicuous water bugs usually found in dense vegetation or on and under stones and debris. The various species are adapted for very different types of aquatic habitats, some occurring in rapid flowing streams while others are found in still ponds and in springs. Five genera occur north of Mexico but three are very rare and local in distribution.

1 Anterior margin of pronotum between and behind eyes straight or only slightly and evenly concave .... *Pelocoris*

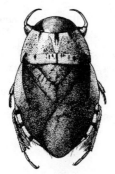

Figure 492

Figure 492 *Pelocoris femoratus* (P.B.)

This is a broadly oval greenish-yellow insect suffused with olive on the front wings. Dead specimens usually fade to dull yellowish brown. The posterior portion of the pronotum is often dull white with dusky markings. The median lobe of the sixth visible ventral segment of the females is not cleft or divided. The length is from 9-11.5 mm. This is a very common species in ponds with abundant vegetation throughout the eastern United States from New England south to Florida and west at least to the Dakotas and Louisiana.

P. carolinensis Torre Bueno is a very similar slightly smaller species found in the southeastern states. Females are readily distinguishable by having the sixth visible abdominal segment deeply cleft on the underside. Males are more difficult to distinguish but the anterior portion of the pronotum of *carolinensis* has more coarse punctures and irregular ridges than does *femoratus*.

A third species *P. shoshone* La Rivers occurs in springs and marshes in Nevada and southern California.

1a Anterior margin of pronotum very deeply concave between and behind eyes. Fig. 493 .... *Ambrysus*

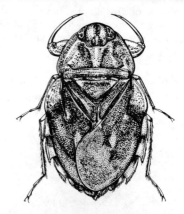

Figure 493

Figure 493 *Ambrysus mormon* Montandon

This is a rather large species (9-12 mm), of a yellowish brown color usually with longitudinal dark markings on the head and pronotum. The front wings are dark brown, frequently uniformly so, but often with yellow mottling present. The anterolateral margin of the front wing (embolium) is greatly expanded and has a somewhat flap-like appearance. This species occurs in flowing streams usually with pebbly bottoms, in springs and marshes and along the margins of lakes which are sometimes of considerable salinity. The distribution is western from California to New Mexico and north to western Nebraska, South Dakota and into southern Montana.

There are eleven additional species of *Ambrysus* in the southwestern United States, most of them with very restricted ranges.

REFERENCES

La Rivers, I. 1951. A revision of the genus *Ambrysus* in the United States. Univ. Calif. Pub. Ent. 8(7):277-338.

La Rivers, I. 1971. Studies of Naucoridae (Hemiptera). Mem. Biol. Soc. Nevada pp. 1-120 (contains a checklist of world species).

GELASTOCORIDAE
The Toad Bugs

This is a very small family of ovoid, flattened dull colored hemipterans that generally live on mud and sand along the margins of ponds and streams. The family is characterized by the prominent bulging eyes, grasping front legs, and antennae concealed below the head. The front and middle tarsi are one-segmented or unsegmented, the hind tarsi are three-segmented. All members of the family are predaceous and feed upon small insects and mites.

black, but it usually is a dull mottled brown and black. The color frequently agrees with the color of the substrate upon which it lives. The somewhat irregular surface and protrudant eyes make the name "toad bug" a very apt one. The legs are usually ringed with dark and white bands. The length is from 7.5-9 mm. The toad bug is usually found along the margins of rivers, streams and ponds. It is found over nearly the entire United States but is scarce or absent in some of the northern states.

One additional species, *G. rotundatus* Champ., is found in the southwest. It can readily be distinguished from *G. oculatus* by its rounded instead of sharp pronotal humeral angles.

1   Front tarsi present and bearing two claws (fig. 494); fore femora not enormously enlarged at base, about twice as long as basal width .... *Gelastocoris*

1a   Front legs without tarsal segments, a single thick claw present arising directly from tip of tibia (fig. 495); fore femora greatly enlarged at the base and about as wide as long .... *Nerthra*

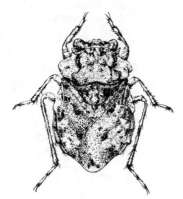

Figure 494

Figure 494 *Gelastocoris oculatus* (F.)

The color of this common toad bug is extremely variable from nearly uniformly yellow to almost

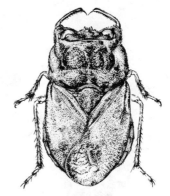

Figure 495

Figure 495 *Nerthra martini* Todd

This is a thick, heavy-bodied insect, 7-9 mm in length, and generally of a brown or reddish-brown color. The wings are separated from one another and a distinct membrane is present. *N. martini* often occurs in muddy situations but some tropical members of the genus live far from water. The distribution is southwestern where it is known from California, Arizona and Nevada.

N. stygica (Say) closely resembles *martini* but has the wings fused together and lacks a distinct membrane in the wing. It is known from Georgia and Florida.

Three additional rare species occur in the United States; one in California, one in Arizona and the third in Florida.

REFERENCE
Todd, E. L. 1955. A Taxonomic Revision of the Family Gelastocoridae. Univ. Kans. Sci. Bull. 37(1):277-475.

OCHTERIDAE
The Velvety Shore Bugs

This is a very small family of oval, black or dark brown often velvety textured insects. Ocelli are present, and the antennae are four-segmented with the first two segments the shortest. The hind tarsi are three-segmented but the front and middle tarsi have only two segments (first segment very small). The front legs are not thickened for grasping prey; the labium is long, usually extending back between the hind legs. These predaceous insects usually live on mud or sand flats, or among grass and weeds along the margins of ponds and streams, or in nearby damp places. Their dull coloration and cryptic habits necessitates special effort to collect them successfully. Only one genus and seven species are known from the United States.

Figure 496 *Ochterus americanus* (Uhler)

In this species the body is dark velvety blue-black and dotted with tiny yellow dots. The pronotum has a large yellow spot on the explanate margin near the antero-lateral angle. Five small yellow spots are usually present along the outer margin of the corium. The legs are yellow. The length is 4 mm. The distribution is from southern New England over much of the eastern United States and west to Texas.

O. banksi Barber is a brownish-black species very similar in appearance to *americanus* but readily recognizable by a small tubercle on the convex portion of the pronotum near the antero-lateral margin. It is distributed from New York to Florida in the eastern United States and southwest at least to central Texas.

REFERENCE
Schell, D. V. 1943. The Ochteridae (Hemiptera) of the Western Hemisphere. J. Kans. Ent. Soc. 16:29-46.

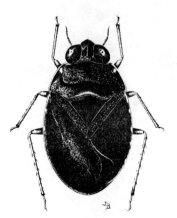

Figure 496

List of Common Names

The following are common names of North American Hemiptera as approved by the Entomological Society of America.

Thyreocoridae

negro bug   *Corimelaena pulicaria* (Germ.)

Pentatomidae

green stink bug	*Acrosternum hilare* (Say)
conchuela	*Pitedia ligata* (Say)
Say stink bug	*Pitedia sayi* (Stål)
western brown stink bug	*Euschistus impunctiventris* (Stål)
brown stink bug	*Euschistus servus* (Say)
dusky stink bug	*Euschistus tristigmus* (Say)
one-spot stink bug	*Euschistus variolarius* (P. de B.)
harlequin bug	*Murgantia histrionica* (Hahn)
southern green stink bug	*Nezara viridula* (L.)
rice stink bug	*Oebalus pugnax* (F.)
two-spotted stink bug	*Perillus bioculatus* (F.)
spined soldier bug	*Podisus maculiventris* (Say)

Coreidae

horned squash bug	*Anasa armigera* (Say)
squash bug	*Anasa tristis* (Deg.)
boxelder bug	*Leptocoris trivittatus* (Say)
leaf-footed bug	*Leptoglossus phyllopus* (L.)

Pyrrhocoridae

Arizona cotton stainer	*Dysdercus mimulus* Hussey
cotton stainer	*Dysdercus suturellus* (H.-S.)

Lygaeidae

hairy chinch bug	*Blissus hirtus* Montd.
chinch bug	*Blissus leucopterus* (Say)
western chinch bug	*Blissus occiduus* Barber
small milkweed bug	*Lygaeus kalmii* Stål
false chinch bug	*Nysius ericae* (Schill.)
large milkweed bug	*Oncopeltus fasciatus* (Dall.)

Reduviidae

wheel bug	*Arilus cristatus* (L.)
masked hunter	*Reduvius personatus* (L.)
western bloodsucking conenose	*Triatoma protracta* (Uhl.)
bloodsucking conenose	*Triatoma sanguisuga* (Lec.)

Cimicidae

bed bug	*Cimex lectularius* L.
poultry bug	*Haematosiphon inodorus* (Dugès)
swallow bug	*Oeciacus vicarius* Horv.

Tingidae

oak lace bug	*Corythucha arcuata* (Say)
hackberry lace bug	*Corythucha celtidis* (O. & D.)
sycamore lace bug	*Corythucha ciliata* (Say)
hawthorn lace bug	*Corythucha cydoniae* (Fitch)
cotton lace bug	*Corythucha gossypii* (F.)
chrysanthemum lace bug	*Corythucha marmorata* (Uhl.)
elm lace bug	*Corythucha ulmi* (O. & D.)
eggplant lace bug	*Gargaphia solani* Heid.
basswood lace bug	*Gargaphia tiliae* (Walsh)
azalea lace bug	*Stephanitis pyrioides* Scott
rhododendron lace bug	*Stephanitis rhododendri* Horv.
lantana lace bug	*Teleonemia scrupulosa* (Stål)

Miridae

alfalfa plant bug	*Adelphocoris lineolatus* (Goeze)
rapid plant bug	*Adelphocoris rapidus* (Say)
superb plant bug	*Adelphocoris superbus* (Uhl.)

ragweed plant bug	*Chlamydatus associatus* (Uhl.)
suckfly	*Cyrtopeltis notatus* Dist.
yucca plant bug	*Halticotoma valida* Reut.
garden fleahopper	*Halticus brachteatus* Say
onion plant bug	*Labopidea allii* Knight
meadow plant bug	*Leptopterna dolabrata* (L.)
caragana plant bug	*Lopidea dakota* Knight
phlox plant bug	*Lopidea davisi* Knight
apple red bug	*Lygidea mendax* Reut.
tarnished plant bug	*Lygus lineolaris* P. de B.
hollyhock plant bug	*Melanotrichus althaeae* (Hussey)
ash plant bug	*Neoborus amoenus* Reut.
hickory plant bug	*Neolygus caryae* Knight
pear plant bug	*Neolygus communis* Knight
four-lined plant bug	*Poecilocapsus lineatus* (F.)
cotton fleahopper	*Psallus seriatus* (Reut.)
hop plant bug	*Taedia hawleyi* (Knight)

Belostomatidae

giant water bug — *Lethocerus americanus* (Leidy)

Gelastocoridae

toad bug — *Gelastocoris oculatus* (F.)

Index and Glossary

A

abbreviatus, Holcostethus, 55
abbreviatus, Phlegyas, 77
abdominalis, Melanolestes, 129
Abedus, 241
ABRADED: scraped, rubbed.
Acalypta, 116
Acanthocephala, 58
Acantholomidea, 43
Acanthophysa, 100-101
Acanthosomatidae, 40
Acanthosomidae, 40, 45
accerra, Thyanta, 51
acuminatum, Desmodium, 60
Acer, 108, 109
acetic acid, 13
Acetropis, 152
Acholla, 123
Acrosternum, 50
Actitocoris, 149
acuminata, Adelia, 112
acuminata, Ramphocorixa, 232
acuminatus, Oncerotrachelus, 120
acutangulus, Thasus, 57
ACUTE: pointed, forming an angle less than 90°.
acutus, Aradus, 104
Adelia, 112
Adelphocoris, 169-70
adjunctus, Cimex, 141
ADPRESSED: to press to something else; press flat.
aeneifrons, Homaemus, 44-45
aequale, Camptozygum, 162
affiguratus, Hoplomachus, 203
Agnocoris, 160
Aknisus, 102
alaskensis, Callicorixa, 236-37
alba, Betula, 41
albescens, Porpomiris, 150
albocinctus, Pachybrachius, 88
albofasciatus, Spanagonicus, 200-201
albonotota, Microvelia, 216
album, Chenopodium, 180
alcohol, 11, 14
Alder, 109, 111
Alepidia, 173
Alepidiella, 173
Alfalfa, 170

Alfalfa Plant Bug, 170
algae, 230
allii, Labopidea, 179
Allorhinocoris, 167
Alnus, 109, 111
alternatus, Archimerus, 60
alternatus, Eurygaster, 43
alternatus, Nabis, 138
althaeae, Melanotrichus, 181
Alveotingis, 113-14
Alydidae, 9, 24, 29, 63
Alydus, 65-66
Amaranthus, 98
Amaurochrous, 40, 42
Amblytylus, 201
Ambrysus, 242-43
Ambush Bugs, 118
americana, Acetropis, 152
americana, Cymatia, 231
americana, Microvelia, 216
americana, Phymata, 119
americanus, Anapus, 172
americanus, Ceanothus, 112
americanus, Ceraleptus, 61
americanus, Conostethus, 200
americanus, Lethocerus, 240
americanus, Monalocoris, 191
americanus, Ochterus, 244
americanus, Plinthisus, 78
americoferus, Nabis, 138
Amnestus, 33-34
amoena, Mesovelia, 217
amoenus, Closterocoris, 194
amoenus, Macrotylus, 201
amoenus, Tropidosteptes, 158
Amorpha, 66
Amphicarpa, 116
amyl acetate, 13
Anapus, 172, 177
Anasa, 62-63
ANASTOMOSING: running together or into each other.
anchorago, Stiretrus, 42
andreae, Dysdercus, 96
Andropogon, 82, 84
Aneurus, 104-5
angulata, Gargaphia, 112
ANGULATE: in the form of an angle.
angulatus, Blepharidopterus, 185
angustatus, Cymus, 81
angustomarginatus, Lygaeus, 71

Ankylotylus, 196
annulata, Emesa, 133
annulata, Pronotacantha, 101
annulatus, Lasiomerus, 137
annulicornis, Rocconota, 126
annulicrus, Phlegyas, 77
ANNULUS: a ring encircling a segment, spot or mark.
ANTENNAE: paired sensory organs located on the head. Fig. 1.
ANTENNAL SOCKET: the membranous area of the head in which each antenna is located.
antennata, Hesperotingis, 114
antennator, Chariesterus, 57
antenniferous tubercle, 3, 4
ANTENNIFEROUS TUBERCLES: raised areas or projections bearing the antennae.
ANTEOCULAR: before the eye.
antevolens, Anthocoris, 146
Anthocoridae, 9, 23, 31, 143, 147, 187
Anthocoris, 146
anthracinus, Ectopiocerus, 156
Antillocoris, 94
Antirrhinum, 101
ant mimetic, 13, 63, 66, 97, 134, 136, 149, 153-55, 171, 193, 221
ants, 172
Apateticus, 47
Aphids, 188
APICAL: the tip or outermost part.
APICAL CORIAL MARGIN: the distal margin of the corium to which the membrane is attached.
apicalis, Stachyocnemus, 65
apiculata, Nepa, 239
Apiomerus, 121-22
Apolygus, 163
APOSEMATIC: colored or constructed so as to indicate special qualities of defense; warning colors.
apple, 109, 178
Apple Red Bug, 161
APPRESSED: closely applied to, pressing against.

approximatus, Pinalitus, 162
aptera, Fitchia, 125
APTEROUS: without wings.
apterus, Halticus, 179
aquatic collecting, 10
aquilinum, Pteridium, 51
Aradidae, 9, 17, 18, 25, 31, 103
Aradus, 103-4
arborea, Brochymena, 48
ARBOREAL: living in or on trees.
Arbor Vitae, 169
arbutifolia, Pyrus, 107
Archimerus, 60
arcuata, Corythucha, 109
ARCUATE: arched, like a bow.
arenaria, Nannocoris, 209
arenarius, Trapezonotus, 92
argenticollis, Gerris, 212
Argyrocoris, 182
Arhaphe, 97
Arhyssus, 70
Arilus, 123
arizonensis, Stenolemoides, 134
armigera, Anasa, 63
AROLIUM: a padlike structure at the base of the tarsal claws.
Artemisia, 107
Ash, 108, 112, 113, 158
Ash-gray Leaf Bugs, 97, 98
aspirator, 7
assassin bugs, 1, 119, 120, 123
associata, Corythucha, 108
associatus, Chlamydatus, 201
Aster, 107, 109, 171, 201
aterrima, Galgupha, 38
Atheas, 115
Atomoscelis, 200
atopodonta, Hesperocorixa, 234
atra, Galgupha, 38, 39
ater, Capsus, 157
Atrachelus, 125-26
Atractotomus, 197
Atrazonotus, 90-91
atricolor, Lepidopsallus, 196
atricornis, Elasmostethus, 42
ATRIUM: any chamber just inside a body opening, such as within a spiracle.
audeni, Callicorixa, 236-37
Aufeius, 67

249

AURICLE: a small lobe or ear-like projection.
aurivilliana, Pseudopamera, 84
Azalea, 115

B

backswimmers, 224, 230
bakeri, Belostoma, 241
balius, Larinocerus, 195
Ballella, 175
bamboo, 110
Banasa, 49-50
banksi, Ochterus, 244
Baptisia, 65, 116
Barber, Harry Gardner, 17
Barberiella, 154
Barce, 132
bark beetles, 143
barn swallow, 140
basalis, Pachybrachius, 88
basalis, Polymerus, 170-71
basicornis, Ballella, 175
basidentatus, Amnestus, 34
basswood, 163, 204
bats, 139, 140, 141
batbugs, 1, 139
BEAK: a pointed sheath formed by the labium.
Beamerella, 195
beans, 112
beating tray, 8
Beckocoris, 196
bedbug, 1, 130, 131, 139, 140
bedstraw, 171, 197
behrensi, Heterogaster, 75
belfragei, Melanopleurus, 72
belfragei, Metatropiphorus, 136
belfragei, Protenor, 64
bella, Saileria, 183
bellula, Alepidia, 173
bellula, Corythaica, 110
Belonochilus, 79
Belostoma, 240-41
Belostomatidae, 21, 27, 239, 241, 242
benzol (benzene), 13
Bergroth, Ewald, 16
Berlese funnel, 8
berries, 1, 39, 49
Berytidae, 24, 29, 99
Berytinidae, 99
Berytinus, 100
Betula, 41, 108, 109
biceps, Systelloderes, 117-18
bicrusis, Neacoryphus, 73
biennis, Oenothera, 102
BIFID: cleft, forked.
bifida, Cenocorixa, 237
Bifidungulus, 178
biflora, Impatiens, 162
biflora, Stylosanthes, 115
"big bed bug", 130
biguttatus, Rasahus, 130
bijugis, Homaemus, 45
bilobatus, Pachybrachius, 88
bimaculata, Cosmopepla, 52
binotatus, Stenotus, 166
bioculatus, Perillus, 1, 46
bipunctata, Tetyra, 44
birch, 41, 81, 95, 108, 109
bispina, Schizoptera, 209
blacklight, 8
black locust, 184
bladder nut, 107
Blatchley, Willis Stanley, 17

Blepharidopterus, 185
Blissus, 78
blue grass, 151, 201
bohemani, Crophius, 79
bohemani, Monosynamma, 202
Bolteria, 167, 169
BOREAL: pertaining to the north: the faunal region extending from the polar sea southward to approximately the northern border of the U.S., but also includes the higher parts of the major mountain ranges.
Bothynotus, 188
bouchervillei, Salda, 222
Bouteloua, 48
Box Elder Bug, 66, 68
Box Elder tree, 68
brachialis, Velia, 216
Brachyceratocoris, 199
Brachynotocoris, 183
BRACHYPTEROUS: with short or abbreviated wings.
bracken fern, 51
bractatus, Halticus, 179
bracteata, Amphicarpa, 116
breviceps, Cymodema, 81
brevipennis, Emesaya, 132
brevirostris, Physatocheila, 111
brevis, Cimex, 141
brevitylus, Amaurochrous, 42
broad-headed bugs, 63
Brochymena, 47
bromeliad, 215
broom, 185
brown stink bug, 55
BUCCULAE: elevated plates or ridges on the underside of the head, one on each side of the beak. Figs. 2, 3, 4.
Buenoa, 224, 225
buenoi, Gerris, 212
buenoi, Palmacorixa, 236
bulbosa, Corythucha, 107
bullatus, Geocoris, 75
burmeisteri, Hebrus, 219
burr oak, 204
burrowing bugs, 33, 37
bush clover, 116
butleri, Pseudometapterus, 132
butterfly pea, 116
buttonbush, 163
butternut, 109

C

cabbage, 53
cacti, 191
calcaratus, Alydus, 66
calcium cyanide, 8
California condor, 142
californica, Graptocorixa, 231
californica, Trichopepla, 54
californicus, Eumops, 139
californicus, Notholopus, 157
californicus, Xyonysius, 80
caligineus, Cyrtocapsus, 192
CALLI: thickened raised spots or swellings on the dorsum of the thorax.
Callicorixa, 236-37
Calliodes, 145
CALLOUSED: with calli.
Calocoris, 164, 170
calva, Banasa, 50

calva, Trichocorixa, 232
Calyptodera, 157
camel's hair brush, 13
campestris, Lyctocoris, 143, 144
Camptozygum, 162
Campylomma, 200
canadensis, Criocoris, 197
canadensis, Pseudocnemodus, 84-85
canadensis, Tetraphleps, 146
canaliculatus, Gerris, 211
canescens, Amorpha, 66
Capsus, 157
cardinalis, Tropidosteptes, 158
Carduus, 108
Carex, 76, 93
caricis, Cyrtorhinus, 186
CARINA: an elevated ridge or keel.
carinata, Galgupha, 38
carinata, Sirthenea, 129
CARINATE: with a keel or ridge.
carolina, Arhaphe, 97
carolina, Ploiaria, 131
carolinensis, Narvesus, 128
carolinensis, Pelocoris, 242
Carpilis, 89
carrion, 34, 61, 63, 66
Carthasis, 135
Carya, 108, 110, 111
caryae, Neolygus, 163
Cassia, 37
Castanea, 109
CASTANEOUS: chestnut brown, bright red-brown.
castaneus, Scaptocoris, 33
caterpillars, 46, 47
Catorhintha, 62
cat-tails, 81
Caulotops, 191
cavernis, Primicimex, 140
caviceps, Hyalochloria, 183
cavifrons, Neottiglossa, 54
Ceanothus, 43, 65, 112
cedar, 173
celery, 162
cellucotton, 8
Cenocorixa, 237
Ceraleptus, 61
Ceratocapsus, 174
Ceratocombidae, 206
Ceratocombus, 206
Ceratopidea, 177
Chaetofoveolocoris, 151
Chaetophylidea, 203, 204
Chaetura, 142
Chagas fever, 130
Chariesterus, 57-58
Chelinidea, 59
Chenopodium, 180
chestnut, 109
chickens, 141, 142
chimney swift, 142
China, William E., 16
chinch bug, 1, 78
Chlamydatus, 201
Chlorochroa, 51
chokeberry, 107
chromosome study, 13
chrysanthemum, 107, 109
cibarium, 5
cigar boxes, 15
ciliata, Corythucha, 108
CILIATE: fringed with a row of parallel hairs.
ciliatus, Cydnoides, 37

ciliatus, Cyrtomenus, 35
Cimex, 140
Cimexopsis, 142
Cimicidae, 18, 31, 139, 143
cimicoides, Macrocephalus, 119
cinctipes, Amaurochrous, 42
cinctus, Largus, 97
cinctus, Psellipous, 125
cinctus, Sehirus, 34
cinerea, Juglans, 109
cinerea, Piesma, 98
cinereus, Atrachelus, 126
cinnamomeus, Aradus, 103
circumcinctus, Perillus, 46
cixiiformis, Myiomma, 205
clandestina, Opistheurista, 156
claripennis, Labopella, 174
CLAVAL SUTURE: the suture or seam between the clavus and corium. Fig. 1.
clavata, Melanorhopala, 114-15
CLAVATE: clubbed, thickened toward the tip.
clavigerus, Sisamnes, 89
CLAVUS: the anal potion of the hemelytra; that part next to the scutellum when folded. Fig. 1.
CLEFT: split, forked.
Clematis, 179
cliff swallow, 140
Clitoria, 116
clitoriae, Leptopharsa, 116
Clivinema, 188
Closterocoris, 194
clypealis, Leptoglossus, 59
CLYPEUS: that part of the head anterior to the front.
Cnemodus, 82
Cnicus, 108
COALESCE: unite, grow together.
Coccobaphes, 159
coelestialium, Trigonotylus, 153
Coenus, 54
Coleoptera, 17, 37, 42
COLEOPTEROID: beetle-like.
COLLAR: the narrow anterior part of the pronotum usually separated from the posterior part by a groove. Fig. 1.
Collaria, 149-50
collaris, Diplozona, 189
collembola, 207
coloradensis, Hesperocimex, 142
Colorado potato beetle, 1, 46
coltsfoot, 165
comatus, Gerris, 212
common bedbug, 141
COMMISSURE: the line of union where the hemelytra meet along the clavus below the apex of the scutellum. Fig. 1.
communis, Neolygus, 163
communis, Tominotus, 36
comosus, Synxenodorus, 142
Compositae, 109, 124, 161, 204
COMPRESSED: flattened laterally.
CONCAVE: hollowed out.
conchuela, 51
concinnus, Hebrus, 219
concolor, Orthotylus, 185

cone-noses, 119
conformis, Gerris, 211
confusa, Buenoa, 226
confusa, Exptochiomera, 90
conicola, Gastrodes, 76
conifers, 59, 146
CONNEXIVUM: the lateral margin of the abdomen at the juncture of the dorsal and ventral sclerites.
Conocephalocoris, 187
Conostethus, 200
consolidus, Hebrus, 219
consors, Hoplomachidea, 198
consors, Maurodactylus, 196
conspersus, Alydus, 66
conspersus, Pachypeltocoris, 170
CONSPURCATE: sprinkled with spots.
constrictus, Perigenes, 86
CONTIGUOUS: touching together.
Coquillettia, 193
Coreidae, 9, 24, 29, 57, 59
CORIACEOUS: leather-like; tough, rigid.
coriaceus, Orthocephalus, 178
coriacipennis, Cymus, 81
Corimelaena, 39, 40
Coriomeris, 60
Corisella, 237
CORIUM: that portion of the hemelytra exclusive of the clavus, membrane, and, if present, the embolium. Fig. 1.
Corixidae, 4, 11, 20, 26, 230
Corixidea, 208
corn, 78
corn earworm, 146
Corticoris, 206
cornuta, Saldoida, 221
Corylus, 109
Corythaica, 110
Corythucha, 107-9, 110, 111, 115
Cosmopepla, 52
COSTAL: pertaining to the costa, the thickened anterior margin of the wing.
costalis, Zerideoneus, 86
costata, Leptoypha, 113
COSTATE: with elevated ridges.
Cotoneaster, 107
cotton, 8, 51, 56, 59, 170
cotton flea hopper, 199
cotton stainer, 95-96
cottonwood, 111
COXA: the basal segment of the leg. Fig. 2.
COXAL CAVITY: the area in which the coxa articulates.
crab apple, 161
crassicornis, Salda, 223
crassimana, Oedancala, 76
crassipes, Apiomerus, 122
crassus, Drymus, 95
Crateagus, 107
creeping water bugs, 242
CRENATE: with small blunt teeth; scalloped.
crenatus, Aradus, 103
Creontiades, 160, 168
Criocoris, 197
cristatus, Arilus, 123

Crophius, 78-79
crow quill pen, 14
cruciata, Rhiginia, 128-29
cruciatus, Elasmostethus, 41
crustacea, 224
Cryphula, 95
CRYPTIC: hidden, concealed.
Cryptostemma, 207
Cryptostemmatidae, 206
cucumber, 63
cucurbit, 63
cucurbitaceus, Dicyphus, 195
cultivated crops, 57
cunea, Hypantria, 46
cuneatus, Neocapsus, 159
CUNEUS: the small triangular area at the end of the embolium in the hemelytra of some Hemiptera.
Curicta, 239
currants, 166
cursitans, Xylocoris, 145
curtipendula, Bouteloua, 48
curtulus, Porpomiris, 150
custator, Thyanta, 52
CUTICULA: the outer covering of an insect.
Cydnidae, 22, 23, 29, 35
Cydnoides, 37
cydoniae, Corythucha, 107
Cyalpus, 186
Cymatia, 231
Cymodema, 81
Cymoninus, 80
Cymus, 81
cynicus, Apateticus, 47
Cyphopelta, 172
cypress, 173
Cyrtocapsus, 192
Cyrtomenus, 35
Cyrtopeltis, 194
Cyrtopeltocoris, 171
Cyrtorhinus, 186

D

Dacerla, 154-55
Dagbertus, 163
dahlias, 166
daisies, 166
daleae, Ceratopidea, 177
Daleapidea, 175
Dallasiellus, 37
damsel bug, 1, 134
davisi, Lopidea, 184
DECLIVENT: sloping downward.
decolor, Corisella, 237
decolor, Lopus, 201-2
decora, Ecertobia, 167
decoratus, Carthasis, 135
DECUMBENT: bending downward.
DEFLEXED: abruptly bent downward.
delicatus, Brachynotocoris, 183
delius, Coenus, 54
demissa, Prunus, 109
DENTICLE: small tooth-like development on a structure.
denticulata, Acantholomidea, 43
Deraeocapsus, 190
Deraeocoris, 190
dermestid beetles, 15
Derophthalma, 157

Desmodium, 60, 65
diadema, Sinea, 124
Diaphnidia, 183, 184
Diaphnocoris, 184
Dichaetocoris, 180
Dichrooscytus, 169
Dictyla, 113
Dicyphus, 194, 195
DIFFUSE: spreading out; without distinct margins.
diffusus, Ligyrocoris, 85
DILATED: widened; expanded.
dimidiata, Banasa, 49
DIMORPHIC: existing in two forms.
Diphleps, 205
Diplozona, 189
Dipsocoridae, 25, 28, 31, 206-7
DISC: the central upper surface of any part.
DISCAL CELL: see fig. 210.
DISCOIDAL: relating to the disc, a rounded plate.
disconotus, Crophius, 79
discors, Cymus, 82
discrepans, Dallasiellus, 37
discolor, Teratocoris, 152
DISJUNCT: separated, standing apart.
dislocatus, Metriorhynchomiris, 167
dissortis, Gerris, 211, 212
DISTAL: that part farthest from the body.
Distant, William Lucas, 16
distincta, Corythucha, 108
distincta, Rhagovelia, 215
distinctus, Merocoris, 60-61
distinguendus, Salignus, 162
DIVERGENT: spreading out from one point.
dock, 170
dog fennel, 171
dogwood, 163
Dolichomiris, 151
Dolichonabis, 137-38
dolobrata, Leptopterna, 136, 151-52
dorsalis, Oedancala, 76
douglasensis, Mesovelia, 217
dragonflies, 1
Drake, Carl John, 17
dredges, 10
Drymus, 94
duckweed, 216
Dysdercus, 95-96

E

eagles, 142
Ecertobia, 167
echii, Dictyla, 113
echinata, Acanthophysa, 101
Echium, 113
ECOTONE: a transitional zone between two distinct communities, habitats or biogeographic areas.
ectoparasites, 20
Ectopiocerus, 156
edulis, Corisella, 237
eggplant, 112
eggplant lace bug, 112
Elasmostethus, 40, 41
Elasmucha, 40, 41

electric light bugs, 239
elegans, Dichrooscytus, 169
ELIPTICAL: oblong-oval.
elkinsi, Gardena, 133
elm, 109, 183
Elmer's Glue-All, 12
EMARGINATE: notched, with a section cut from a margin.
Emblethis, 94
EMBOLIUM: the differentiated costal part of the corium in the hemelytra of some hemiptera.
Emesa, 133
Emesaya, 131, 132
Emesinae, 99
emoritura, Teratodia, 205
Empicoris, 133, 134
Enicocephalidae, 22, 31, 117
ephedrae, Merinocapsus, 196
EPIMERON: the posterior division of a thoracic pleuron.
EPISTERNUM: the anterior division of a thoracic pleuron.
Eremocoris, 92
ericae, Nysius, 80
eriogoni, Nicholia, 202
Eriogonum, 110, 113
errabundus, Empicoris, 133
ethanol, 13
ethyl acetate, 8, 13
eumops, Hesperoctenes, 139
Eupatorium, 107
Euphorbia, 57, 176
eurinus, Alydus, 66
Europiella, 199
Eurychilopterella, 189
Eurygaster, 43
Euschistus, 46, 55
Eustictus, 188, 189
Euthochtha, 61
EVAPORATIVE AREA: differentiated area around scent gland opening. Fig. 6.
exaptus, Perillus, 47
EXPLANATE: flattened; applied to a margin.
Exptochiomera, 90
exsangius, Zelus, 124
extensa, Corimelaena, 40
extensum, Xenogenus, 67
eye, 3

F

FACET: one of the lens-like divisions of the compound eye.
Fagus, 105
FALCATE: sickle-shaped; convexly curved.
fall webworms, 1, 46
falicus, Ischnodemus, 77
fallax, Metriorhynchomiris, 167
false chinch bug, 80
false Solomon's seal, 167
farfara, Tussilago, 165
FASCIA: a transverse band or broad line.
fasciatus, Oncopeltus, 71
femorata, Acanthocephala, 58
femoratus, Pelocoris, 242
FEMUR: usually the stoutest segment of the leg; that segment between the trochanter and tibia. Fig. 1.

Index and Glossary 251

ferns, 191
ferrugata, Leptopterna, 152
ferus, Eremocoris, 93
ferus, Nabis, 138
FILIFORM: thread-like.
filiformis, Ganocapsus, 167
filiformis, Leptocorisa, 64
fingernail polish, 13
fiskei, Aneurus, 105
Fitchia, 125
flat bugs, 103-5
flavicornis, Scoloposcelis, 143
flavoscuta, Lopidella, 175
flavosparsus, Melanotrichus, 180
fleabane, 163
flower bugs, 1
flumineum, Belostoma, 240, 241
food canal, 5
FORBES: nongrassy herbaceous vegetation.
formicina, Hymenocoris, 117
FOSSA: a pit or deep sulcus.
FOSSORIAL: formed for digging or burrowing.
fourlined leaf bug, 166
FRACTURE: the indentation in the hemelytra separating the cuneus from the corium.
franseriae, Stittocapsus, 168
fraterna, Barce, 132
fraternus, Peritrechus, 91
fraternus, Prepops, 156
Fraxinus, 108, 112, 113
free-tailed bat, 139
FRONS: the front part of the head anterior to the vertex. Fig. 4.
Fulvius, 187
fumidus, Noctuocoris, 180
fungi, 186
fungus beetles, 143
fusca, Pagasa, 135
fusca, Ranatra, 238
FUSCOUS: dark brown, a mixture of black and red.
fuscovittata, Saica, 121
fusculus, Lasiochilus, 144
fusiformis, Garganus, 165

G

gagates, Mallochiola, 147
galactinus, Xylocoris, 145
galeator, Euthochtha, 60, 61
Galeatus, 106
Galgupha, 38, 39, 40
Galium, 197
Ganocapsus, 167, 170
Gardena, 133, 134
garden flea hoppers, 179
Garganus, 165
Gargaphia, 111, 112
Gastrodes, 76
Gaura, 102
Gelastocoridae, 11, 20, 27, 243
Gelastocoris, 243
gelastops, Hesperolabops, 191
"Gelva," 13
geniculatus, Oncocephalus, 128
Geocoris, 74
Gerridae, 10, 21, 28, 210
Gerris, 210, 211, 212, 214
Ghinallelia, 132
giant water bug, 239

gillettii, Corimelaena, 40
GLABROUS: smooth, hairless; without punctures.
glacial acetic acid, 13
Globiceps, 182
GLOBOSE: spherical.
glycerin, 13
Glyptocombus, 208
golden rod, 10, 61, 79, 171, 175, 181, 184, 204
good luck bugs, 210
gossypii, Corythucha, 109
gracilis, Alepidia, 173
GRANULATE: covered with small grains.
grape vine, 154, 188
Graptocorixa, 231
griseolus, Myochroocoris, 202
GRISEUS: light gray.
griseus, Lethocerus, 240
grossocerata, Alveotingis, 114
Guanabarea, 188
guildini, Piezodorus, 49
GULA: the sclerite forming the central part of the head beneath. Fig. 3.
guttulatus, Oncotylus, 202
guttulatus, Paraxenetus, 154

H

Hadronema, 185
haematoloma, Jadera, 68
Haematosiphon, 142
haemoglobin, 225
Halobates, 210
Halticotoma, 191
Halticus, 178-79
Hammatocerus, 121
harlequin bug, 1, 53
Harmostes, 67
hawthorn, 107, 161, 178
hazel, 109
Hebridae, 21, 25, 27, 218, 219
hebroides, Merragata, 219
Hebrus, 10, 219
heidemanni, Alepidiella, 173
heidemanni, Hesperophylum, 187
heidemanni, Leptopharsa, 116
heidemanni, Sericophanes, 171-72
heidemanni, Triatoma, 130
Heidemanniella, 194
HEMELYTRA: the partially thickened, partially membranous front wings of Hemiptera.
hemipterologists, 16
hemipterus, Cimex, 141
Hemisphaerodella, 191
Heraeus, 85
hermsi, Hesperoctenes, 139
hesione, Limnogonus, 212-13
hesperius, Ceratocombus, 206
hesperius, Labops, 182
hesperius, Metrobates, 213
Hesperocapsus, 177, 180
Hesperocimex, 142
Hesperocorixa, 233, 234, 235, 236
Hesperoctenes, 139
Hesperolabops, 191
Hesperotingis, 114
Heterocordylus, 178
Heterogaster, 75

Heterotoma, 175
Hibiscus, 96
hickory, 108, 110, 111, 163
hilare, Acrosternum, 50
hinei, Microvelia, 216
hirsuta, Chaetofoveolocoris, 151
hirticornis, ploiaria, 131
hirtum, Semium, 176
hirtus, Blissus, 78
hirtus, Labops, 183
histrio, Deraeocoris, 190
histrionica, Murgantia, 53
HOLARCTIC: a biogeographic region encompassing North America, Europe, and Northern Asia.
Holcostethus, 55
hollyhock, 108
hollyhock plant bug, 181
Homaemus, 44-45
Homoptera, 2, 205
honey locust, 173
Hoplistoscelis, 136-37
Hoplomachidea, 196-97, 198
Hoplomachus, 202-203
horni, Macrovelia, 218
Horvath, Geza, 16
howardi, Curicta, 239
human bed bug, 139
HUMERI: the posterior lateral angle ("corners") of the pronotum.
humilis, Coriomeris, 60
humilis, Saldula, 224
Hussey, Roland F., 17
Hungerford, Herbert Baker, 17
HYALINE: transparent or partly so; translucent.
hyalinus, Liorhyssus, 69
Hyalochloria, 183
Hyaloides, 188
Hydrometra, 209
Hydrometridae, 10, 21, 27, 99, 209
Hymenocoris, 117
Hypantria, 46
Hypogeocoris, 74

I

Icodema, 204
Ilnacora, 181-82
Ilnacorella, 181
immaculatus, Abedus, 241
Impatiens, 162
IMPRESSED: having shallow depressed or indented areas.
impressicollis, Aufeius, 67
IMPUNCTATE: not marked with punctures.
incognita, Corimelaena, 40
incognitus, Gerris, 212
INCRASSATE: thickened, swollen.
indentatus, Abedus, 241
indica, Notonecta, 228, 229
infirma, Pnirontis, 127
inodorus, Haematosiphon, 142
insidiosus, Orius, 145, 146
insignis, Atheas, 115
insignis, Mimoceps, 154
insperatus, Gerris, 212
insularis, Blissus, 78
insulata, Notonecta, 226, 227
intermedius, Halticus, 179

INTEROCULAR: area between the eyes.
interrupta, Hesperocorixa, 236
Ioscytus, 223
Irbisia, 157, 160, 168
irrorata, Notonecta, 227
irrorata, Reuteria, 183
Ischnodemus, 77
Isometopidae, 9, 23, 31, 205
isopropyl alcohol, 13

J

Jadera, 68
Jalysus, 101
japonica, Pieris, 116
Jersey tea, 112
jewelers forceps, 13
JUGA: the lateral lobes of the head, one on each side of the tylus (singular-jugum). Figs. 2, 3, 4.
juglandis, Corythucha, 109
Juglans, 109
jumping tree bugs, 205
Juniperus, 104, 173

K

kalmii, Lygaeus, 71
kansa, Trichocorixa, 233
KEEL: an elevated ridge or carina.
Keltonia, 198
kennicottii, Hesperocorixa, 234
Kentucky Blue Grass, 201
killing bottles, 8
kirbyi, Notonecta, 227
Kirkaldy, George Willis, 16
kirkaldyi, Ranatra, 238
kissing bugs, 119
Kleidocerys, 79, 80
Klopicoris, 190
knapsack, 8
Knight, Harry Hazelton, 17
Kolenetrus, 82, 83

L

LABIUM: the segmented sheath enclosing the stylets. Fig. 4.
Labopella, 174
Labopidea, 179, 185
Labops, 182-83
LABRUM: the upper "lip," in Hemiptera typically an elongate triangular structure covering the basal part of the labial "groove."
laevigata, Hesperocorixa, 235
laeviscutatus, Rhasis, 159
Lagenaria, 102
Lampethusa, 164
Lamprocanthia, 223
LANCEOLATE: spear-shaped; oblong and tapering to the end; lance-shaped.
lanipes, Stenolemus, 134
larch, 173
large milkweed bug, 71
Largid Bugs, 96
Largidae, 9, 24, 30, 96, 97
Largidea, 188

Largus, 96
Larinocerus, 195
Lasiochilus, 144
Lasiomerus, 137
LATERAD: toward the side and away from the median line.
lateralis, Arhyssus, 70
lateralis, Corimelaena, 40
lateralis, Elasmucha, 41
lateralis, Neacoryphus, 73
laticephalus, Beckocoris, 196
latipennis, Ceratocombus, 206
latipennis, Tetraphleps, 146
leafhoppers, 2
lectularius, Cimex, 141
legumes, 161, 170, 179
Lepidopsallus, 196
Lepidoptera, 42
Leptocoris, 68
Leptoglossus, 58
Leptopodidae, 25, 28, 219
Leptopterna, 136, 151
Lethocerus, 240
leucopterus, Blissus, 78
lichens, 205
ligata, Pentacora, 222
ligata, Pitedia, 51
light trap, 8
Ligyrocoris, 85
limbolarius, Holcostethus, 55
limnocastoris, Buenoa, 226
Limnogonus, 212
linden, 163
linearis, Dolichomiris, 151
lineatus, Poecilocapsus, 166-67
lineolaris, Lygus, 161
lineolatus, Adelphocoris, 170
Liorhyssus, 69
lobata, Hesperocorixa, 235
lobata, Mezira, 106
longiceps, Hesperoctenes, 139
longulus, Paromius, 87
LORUM: the plate or sclerite on the outer sides of the juga and in front of the eyes.
louisianae, Trichocorixa, 233
louisianica, Niesthrea, 69
lucida, Hesperocorixa, 235
lugens, Mormidea, 56
lugubris, Aradus, 104
lugubris, Dallasiellus, 37
lunata, Notonecta, 228, 229
LUNATE: crescent-shaped.
lutarium, Belostoma, 241
luteolus, Xylastodoris, 98
Lyctocoris, 143
Lygaeidae, 17, 23, 24, 28, 30, 70
Lygaeospilus, 72
Lygaeus, 71
Lygidea, 160-61
Lygocoris, 162-63
Lygus, 161
Lygus-bugs, 161

M

Macrocephalus, 119
macroeps, Trichocorixa, 233
Macrolophus, 194
MACROPTEROUS: long-winged
macrotibialis, Buenoa, 225, 226
macrotis, Tarida, 139
Macrotyloides, 181
Macrotylus, 201

Macrovelia, 217
maculiventris, Podisus, 46
maerkelii, Pithanus, 153
major, Corixidea, 208
major, Mecidea, 48
Malacosoma, 46
malina, Ilnacora, 181-82
malinus, Heterocordylus, 178
mammals, 130, 141
man, 141, 144
maple, 95
margaritacea, Buenoa, 225
marginella, Corimelaena, 40
marsh grass, 154
marsh treaders, 209
Martarega, 229
martini, Hydrometra, 209
martini, Nerthra, 243, 244
masked bed bug hunter, 131
Maurodactylus, 140
mavortius, Cnemodus, 82
McAtee, Waldo Lee, 17
Meadorus, 41
meadow plant bug, 136, 151-52
Mecidea, 48
Mecomma, 186
media, Lopidea, 184
Megaloceroea, 152
Megalocoleus, 202
Megalopsallus, 198
Megalotomus, 65
meilleurii, Collaria, 149-50
Melanaethus, 36
Melanolestes, 129
Melanopleurus, 72
Melanotrichus, 180-81
MEMBRANE: any thin transparent body tissue; specifically the transparent or translucent tips of the hemelytra.
mendax, Lygidea, 161
mendica, Catorhintha, 62
Merinocapsus, 196
meriopterum, Heteroma, 175
Merocoris, 60
Merragata, 10, 219
MESEPIMERON: the posterior of the two lateral sclerites making up the mesopleuron. Fig. 6.
MESEPISTERNUM: the anterior of the two lateral sclerites making up the mesopleuron. Fig. 6.
MESOPLEURON: the lateral area of the mesothorax. Fig. 2.
MESOSCUTUM: the anterior part of the mesothorax, lying under the edge of the prothorax, ordinarily not visible.
MESOSTERNUM: the underside of the mesothorax. Fig. 6.
MESOTHORAX: the second thoracic segment bearing the first pair of wings and the middle pair of legs.
Mesovelia, 217
Mesoveliidae, 27, 217
mesquite, 62
METAPLEURON: the lateral area of the metathorax. Figs. 2, 6.
METASTERNUM: the underside of the metathorax. Fig. 6.

METATHORAX: the third thoracic segment (see metapleuron, fig. 2)
Metatropiphorus, 135
METAXYPHUS: a triangular plate lying behind the hind coxae.
Metriorhynchomiris, 167
Metrobates, 213, 214
mexicana, Martarega, 229
mexicana, Tarida, 140
Mexican free-tailed bat, 140
Mezira, 105
Micracanthia, 224
microhabitat, 9, 10
Microphysidae, 23, 31, 147
Microphylellus, 203, 204
Microphylidea, 200
Microporus, 36
MICROPTEROUS: small-winged.
Microvelia, 10, 25
militaris, Hadronema, 185
milkweeds, 71
MIMETIC: imitative, resembling another organism; antmimetic—resembling an ant.
Mimoceps, 153-54
mimulus, Dysdercus, 96
mimulus, Ochrimnus, 72
Mineocapsus, 200
minimus, Mineocapsus, 200
minor, Berytinus, 100
minor, Hesperocorixa, 235
minor, Mecidea, 48
Minorella, Hesperocorixa, 234
mint, 34
minuten nadeln, 13
minute pirate bugs, 1, 143
minutus, Antillocoris, 94
minutus, Rheumatobates, 214
Mirabilis, 62
mirabilis, Hemisphaerodella, 191
Miridae, 9, 11, 17, 23, 31, 147, 205
Mirini, 172, 194
Miris, 168
mississippiensis, Scoloposcelis, 143
mites, 145
modesta, Cyphopelta, 172
modestus, Atomoscelis, 200
modestus, Podisus, 47
moerens, Chaetophylidea, 204
moerens, Ranzovius, 197
mollicula, Megalocoleus, 202
Monalocoris, 191
Monosynamma, 202
Mormidea, 56
mormon, Ambrysus, 242
morphology, 3
moss, 10
mosquito larvae, 224, 230
moths, 1
mounting, 12
mouth parts, 4
Mozena, 61-62
mud flats, 11
mullein, 200
mulsanti, Mesovelia, 217
multispinosa, Acholla, 123
multispinus, Aknisus, 102
Murgantia, 52
musculus, Anthocoris, 146
musette bag, 8
muticus, Neides, 100

Myiomma, 205
Myochroocoris, 202
Myodocha, 82

N

Nabicula, 136
Nabidae, 9, 22, 23, 31, 134
Nabis, 138
Nannocoris, 209
Narvesus, 128
nasicus, Conocephalocoris, 187
nasutus, Amblytylus, 201
Naucoridae, 11, 21, 27, 242
Neacoryphus, 73
nebularis, Gerris, 211
nebulosus, Deraeocoris, 190
nebulosus, Sphragisticus, 92
Needham scraper, 10
negro bugs, 37, 39
Neides, 99
Neididae, 99
Neoborella, 158
Neoborops, 160
Neocapsus, 159
Neolygus, 162
Neoplea, 230
neoprene stoppers, 14
NEOTROPICAL: that portion of the New World extending southward from the Tropic of Cancer.
Neottiglossa, 53
Nepa, 238
Nepidae, 20, 27, 238
Nerthra, 11, 243
nets, 7, 9
nettles, 34
Neurocolpus, 163
Neuroctenus, 105
nevadensis, Brachyceratocoris, 199
Nevadocoris, 197
Nezara, 50
Nicholia, 202
Niesthrea, 69
niger, Nysius, 80
nigra, Corimelaena, 40
nigra, Ranatra, 240
nigristernum, Arhyssus, 70
nigrolineata, Icodema, 204
nitens, Xestocoris, 90
nitida, Hesperocorixa, 235
nitiduloides, Galgupha, 39
NODAL FURROW: a membranous area or "joint" on the anterior edge of the hemelytron in Corixidae.
notabilis, Cymoninus, 80
notabilis, Gerris, 212
Notholopus, 157
Notonecta, 226
Notonectidae, 11, 20, 27, 224, 230
Noctuocoris, 180
nodosa, Ptochiomera, 89
nubilus, Neurocolpus, 163
numenius, Belonochilus, 79
nyctaginea, Mirabilis, 62
nyctalis, Cimexopsis, 142
Nysius, 66, 69, 79, 80

O

oak, 163, 173, 190
oats, 153

obesa, Mozena, 62
obesa, Rhagovelia, 215
obliqua, Hesperocorixa, 235
obliquus, Microporus, 36
OBSOLETE: almost or entirely absent; indistinct.
obtusa, Mozena, 62
OBTUSE: not pointed; of a greater than a right angle.
occidentalis, Leptoglossus, 59
occidentalis, Scoloposcelis, 143
OCELLI: simple bead-like eyes located on the vertex of the head. Fig. 1.
Ochrimnus, 72
Ochteridae, 11, 21, 244
Ochterus, 244
oculata, Collaria, 150
oculatus, Gelastocoris, 243
odiferous orifice, 4
Oebalus, 56
Oeciacus, 140
Oedancala, 76
Oenothera, 102
oleosa, Collaria, 150
OMMATIDIUM: one of the visual elements that compose the compound eye; see facet.
omnivagus, Neolygus, 163
OMPHALIUM: scent gland opening in the middle of the metasternum (Gerridae).
Oncerometopus, 155
Oncerotrachelus, 120
Oncocephalus, 128
Oncopeltus, 71
Oncotylus, 202
Opistheurista, 156
oppositus, Leptoglossus, 59
Opuntia, 59
orange hawkweed, 100
oranges, 96
Oravelia, 218
Orectochilus, 193
oriander, Rhagovelia, 215
Orius, 145
ornatus, Reuteroscopus, 198
Ornithocoris, 142
Orthocephalus, 177
Ortholomus, 79
orthoneuron, Empicoris, 133
Orthops, 161
Orthoptera, 17
Orthotylus, 181, 185
ostracods, 209
ovalis, Galgupha, 39
Owl, 142
Oxytropis, 65
Ozophora, 93

P

pabulinus, Lygocoris, 162-63
Pachybrachius, 87-88
Pachypeltocoris, 170
padi, Corythucha, 109
Pagasa, 134-35
PALA: the shovel-like anterior tarsal segment of the front leg in Corixidae.
PALEARCTIC: a biogeographic region comprising the northern and temperate portions of Eurasia.
Pallacocoris, 165
pallens, Geocoris, 75
pallidovirens, Thyanta, 51
pallidus, Amnestus, 34
pallidus, Ornithocoris, 142
pallidulus, Lasiochilus, 144
pallidulus, Taylorilygus, 163
pallipes, Ankylotylus, 196
pallipes, Corythucha, 108
pallipes, Saldula, 224
Palmacorixa, 236
Pamillia, 174
Pangaeus, 34-35
Panicum, 102, 110
Paradacerla, 155
PARANOTUM: lateral expansions of the sides of the pronotum. Particularly in Tingidae.
Paraplea, 230
Paraproba, 183
Paravelia, 10, 215, 216
Paraxenetus, 154
PAREMPODIA: structures arising between the tarsal claws (=arolia of Miridae).
Paromius, 86-87
Parshley, Howard Madison, 17
Parthenicus, 182
parva, Brochymena, 47
parvulus, Homaemus, 45
Patapius, 220
pear, 163
pectinata, Spartina, 77
pectoralis, Pygolampis, 127
PEDUNCULATE: set on a stalk; petiolate.
pege, Oravelia, 218
PELAGIC: oceanic, inhabiting the sea.
pelagica, Chaetura, 142
pellucida, Diaphnocoris, 184
Pelocoris, 242
pencil flower, 115
pennsylvanica, Phymata, 118-19
pennsylvanicum, Acrosternum, 51
pennsylvanicus, Melanaethus, 37
Pentacora, 221-22
Pentatomidae, 2, 9, 22, 28, 40
pergandei, Corythucha, 109
Perigenes, 86
Perillus, 1, 45-46
Peritrechus, 91, 92
Peritropis, 187, 205
persica, Prunus, 108
persimilis, Pitedia, 51
personatus, Beamerella, 195
personatus, Reduvius, 129, 131
pestifer, Salsola, 110
Petalostemum, 115
petrunkevitchi, Notonecta, 228
Phlegyas, 76
phlox, 184
phorodendronae, Klopicoris, 190
Phylinae, 176
Phyllopidea, 200
phyllopus, Leptoglossus, 59
Phymata, 118-19
Phymatidae, 9, 22, 30, 118
Phymatopsallus, 198
Physatochelia, 110-11
Phytocoris, 165-66
PHYTOPHAGOUS: feeding upon plants.
Picea, 104
piceus, Hypogeocoris, 74
picipes, Melanolestes, 129
picturata, Ozophora, 93-94
pictus, Trepobates, 214
Pieris, 116
Piesma, 98
Piesmatidae, 25, 30, 97
Piezodorus, 49
pigmy backswimmers, 230
pigweed, 98, 180
Pilophoropsis, 172
Pilophorus, 173
PILOSE: covered with soft down or hair.
pilosellus, Cimex, 141
pilosulus, Alydus, 66
pilosulus, Antillocoris, 94
Pinalitus, 162
Pinus, 44, 79, 104, 111, 173
Pitedia, 51
Pithanus, 153
placidus, Podisus, 1, 46
Plagiognathus, 204
plana, Leptodictya, 110
plant bugs, 147
plaster of Paris, 8
Platanus, 108
Platylygus, 159
Plea, 230
plebejus, Heraeus, 85
Pleidae, 11, 20, 26, 230
plenus, Kolenetrus, 83
Plesiocoris, 160
Plesiodema, 204
PLEURAL: pertaining to the lateral surfaces of the body segments.
plexa, Physatochelia, 110-11
Plinthisus, 75, 78
Ploiaria, 131, 134
PLUMOSE: feathered, like a plume.
Pnirontis, 127
Podisus, 46-47
Podopidae, 40
podopioides, Prionosoma, 54
Poecilocapsus, 166-67
point punch, 12
poison hemlock, 162
politus, Ioscytus, 223
politus, Plagiognathus, 204
Polyctenidae, 20, 26, 139
Polymerus, 164, 170-71
POLYMORPHIC: occurring in several forms.
pond skaters, 210
pooter, 7
poplar, 108
poppaea, Gardena, 133
Populus, 108, 111
Porpomiris, 150
PORRECT: extending forward horizontally.
POSTNODAL: in Corixidae the area distal of, or posterior to, the nodal furrow (see nodal furrow).
POSTOCULAR SPACE: longitudinal distance measured from the posterior margin of the eye to the base of the head.
prairie clover, 115
praying mantid, 238
prehensilis, Macrocephalus, 119
Prepops, 155-56
prickly pear cactus, 59
Primicimex, 139-40
Prionosoma, 54-55
Proba, 159
procaine tubes, 14
PRODUCED: extended; drawn out.
productilis, Ghinallelia, 132
propinquus, Dolichonabus, 138
Pronotocantha, 101
Pronotocrepis, 195
PRONOTUM: the upper part of the first thoracic segment. Fig. 2.
PROPLEURON: the lateral area of the prothorax. Fig. 2.
PROSTERNUM: the ventral part of the first thoracic segment. Fig. 6.
PROTHORAX: the first thoracic segment, bearing the anterior pair of legs but no wings.
Protenor, 64
provancheri, Salda, 222
PROXIMAL: that part of an appendage nearest the body.
PRUINOSE: covered with fine dust or "bloom"; as if frosted.
pruni, Corythucha, 109
Prunus, 108, 109
Psallus, 199
Pselliopus, 124-25
PSEUDAROLIA: false arolia; paired structures found beneath the claws in some Hemiptera. pseudoarolia = empodia.
Pseudatomoscelis, 198-99
Pseudocnemodus, 84-85
Pseudometapterus, 132
Pseudopamera, 84
Pseudopsallus, 177
Pseudoxenetus, 172-73
Pteridium, 51
Ptochiomera, 88-89
PUBESCENCE: short fine soft hair or down.
pubescens, Ceanothus, 43
pugnax, Oebalus, 56
pulchella, Microvelia, 215, 216
pulchellus, Corticoris, 206
pulicaria, Corimelaena, 39, 40
pulverulentus, Agnocoris, 160
PULVILLUS: a soft padlike structure between the lateral claws.
pumpkins, 63
PUNCTATE: set with impressed points or punctures.
punctipes, Geocoris, 74-75
punctiventris, Stictopleurus, 69
purcis, Hammacerus, 121
purple martin, 142
purpurem, Petalostemum, 115
pusillus, Amnestus, 33, 34
Pycnocoris, 157
Pycnoderes, 192
Pygolampis, 127
pyrioides, Stephanitis, 115
Pyrrhocoridae, 9, 17, 24, 30, 95
Pyrus, 107

Q

quack grass, 152
QUADRATE: square or nearly so.
quadrilineatus, Aradus, 103
quadrimaculata, Pycnoderes, 192, 193

quadrimaculata, Saldula, 224
quadripustulata, Brochymena, 47
quercalbae, Neolygus, 163
Quercus, 105, 109, 110
quince, 107, 161
quinquespinosus, Megalotomus, 65

R

racemosa, Aralia, 42
ragweed, 171, 198, 201, 204
Ramphocorixa, 232
Ranatra, 238, 239
Ranzovius, 197
Rapidograph, 14
Rapid Plant Bug, 170
rapidus, Adelphocoris, 169, 170
RAPTORIAL: adapted for seizing prey.
raspberries, 195
recticornis, Megaloceroea, 152
RECURVED: to curve or bend back or backward.
redbud, 167
Red Bugs, 95, 96
red cedar, 169
red clover, 65
red maple, 159
reducta, Mezira, 106
Reduviidae, 2, 9, 21, 22, 30, 119
Reduvioidea, 118
REFLEXED: bent back.
reflexulus, Harmostes, 67
regalis, Pseudoxenetus, 173
relaxing fluid, 13
remigis, Gerris, 212
Renodaeus, 172
repetita, Anasa, 63
Repipta, 126
repletus, Dichrooscytus, 169
resedae, Kleidocerys, 81
RESPIRATORY FILAMENTS: adaptations that facilitate respiration, located on the caudal end of some aquatic insects.
reticulata, Ploiaria, 131
RETICULATE: net-like covered with a net-work of lines.
Reuter, Odo Morannal, 17
Reuteria, 183
Reuteroscopus, 198
Rhagovelia, 10, 214, 215
Rhasis, 159
Rheumatobates, 213
Rhiginia, 128
Rhinacloa, 196
Rhinocapsus, 203
rhododendri, Stephanitis, 115
Rhododendron, 115
Rhopalidae, 24, 29, 66
Rhus, 65
Rhynocoris, 122
rice, 56, 64
rice stink bug, 56
riffle bugs, 214
rileyi, Rheumatobates, 213
robiniae, Lopidea, 184
robusta, Calyptodera, 157
robustus, Aradus, 103, 104
robustus, Melanaethus, 37
Rocconota, 126
roots, 1
Rosa, 109

rose, 109
roseipennis, Nabis, 138
ROSTRATE: having a rostrum.
ROSTRUM: see beak.
rotundatus, Gelastocoris, 243
royal palm, 98
royal palm bugs, 98
rubidus, Lepidopsallus, 196
rubromaculatus, Empicoris, 133
rugicollis, Plesiocoris, 160
RUGOSE: wrinkled.
RUGULOSE: minutely wrinkled.
rushes, 81

S

Saica, 120
Saileria, 183
Salda, 222
Saldidae, 25, 28, 219, 220
Saldoida, 220
Saldula, 223, 224
saliens, Criocoris, 197
Salignus, 162
Salix, 108, 110, 111
Salsola, 110
saltator, Glyptocombus, 208
saltator, Orthocephalus, 178
SALTATORIAL: adapted for jumping.
salt flats, 233
sanguinareus, Coccobaphes, 159
sanguisuga, Triatoma, 130
sarcobati, Tannerocoris, 197
Sarothamnus, 185
saxatalis, Aeronautes, 142
Say, Thomas, 17
sayi, Pitedia, 51
Say's stink bug, 51
scale insects, 2
Scalponotatus, 175
Scaptocoris, 33
SCENT GLAND OPENING: the orifice through which the secretion of the internal scent gland escapes onto the evaporative area. Fig. 6.
Scent-less Plant Bugs, 66-70
Schaffneria, 174
Schizoptera, 208
Schizopteridae, 23, 31, 206, 207
"Schmidt" box, 15
scimitra, Buenoa, 225
Scirpus, 76
SCLEROTIZED: hardened in definite areas.
scolopax, Ortholomus, 79
Scoloposcelis, 143
Scolopostethus, 93
scoparius, Andropogon, 82
scoparius, Sarothamnus, 185
scorpions, 7
scrupeus, Taedia, 164
scurrilis, Argyrocoris, 182
scutellaris, Heidemanniella, 194
scutellatus, Orthops, 161, 162
scutellatus, Pseudoxenetus, 172, 173
Scutelleridae, 40
SCUTELLUM: the triangular part of the mesothorax, usually between the bases of the hemelytra, but in some overlapping them. Fig. 1.

seed bugs, 70
SEGMENT: a subdivision of the body or of an appendage between areas of flexibility.
Sehirus, 34
semilucida, Hesperocorixa, 234
Semium, 175, 176
semiustus, Maurodactylus, 196
semivittata, Trichopepla, 54
separators, 8
seriatus, Pseudatomoscelis, 199
sericea, Plesiodema, 204
sericea, Pygolampis, 127
SERICEUS: silky; with short thick silky down.
Sericophanes, 171
serieventris, Podisus, 46
serotina, Prunus, 108, 109
SERRATE: saw-like.
serratus, Harmostes, 67
serripes, Myodocha, 82, 85
servus euschistoides, Euschistus, 56
servus, Euschistus, 56
servus servus, Euschistus, 56
SETA: slender, hair-like structure.
SETIFORM: bristle or seta-shaped.
sexcincta, Trichocorixa, 233
shell vial, 14
shield bugs, 44, 45
shore bugs, 11, 220, 222, 224
shoshone, Pelocoris, 242
sidae, Niesthrea, 69
sifters, 8
Sigara, 237
signatus, Actitocoris, 149
similis, Aradus, 104
similis, Perigenes, 86
similis, Ploiaria, 131
simplex, Aneurus, 105
simplex, Neuroctenus, 105
simulans, Capsus, 157
simulans, Leptodictya, 110
Sinea, 124
sinipes, Sinea, 124
SINUATE: wavy, specifically of an edge or margin.
Sirthenea, 129
Sisamnes, 89
Sixeonotus, 192
Slaterocoris, 174, 175
slime flux, 145
slossoni, Saldoida, 221
"Smaller Milkweed Bug," 71
smaller water striders, 214
smart weed, 190
sobrinus, Hebrus, 219
sodium cyanide, 8
Solanaceae, 112
solani, Gargaphia, 112
Solidago, 109, 115
sonoriensis, Hesperocimex, 142
sordidus, Hoplistoscelis, 137
sorghum, 56
soybeans, 65
Spanish moss, 209
Spanagonicus, 200, 201
Spartina, 153
SPERMALEGE: a unique organ or group of cells modified to receive sperm; found in the female abdomen of certain Hemiptera exhibiting traumatic insemination.
Sphaerobius, 83
Sphragisticus, 91

spider webs, 132, 134
spikenard, 42
spinifrons, Amnestus, 34
spinifrons, Galeatus, 106, 107
spinosus, Jalysus, 102
spinosus, Patapius, 220
SPINULETS: small spines.
spinulosa, Fitchia, 125
spinulosa, Stenopoda, 127
SPIRACLE: a breathing pore. Fig. 1.
Spiraea, 81
spissipes, Apiomerus, 122
springtail, 145
Squamocoris, 177
squash, 63
Squash Bug, 1, 57, 63
squirt gun, 11
Stachyocnemus, 64
stagnalis, Paravelia, 216
Stål, Carl, 17, 18
stali, Lyctocoris, 144
Staphylea, 107
St. Augustine grass, 78
stems, 1
Stenodema, 150
Stenolemoides, 134
Stenolemus, 133
Stenopoda, 127
Stenotus, 166
step block, 14
Stephanitis, 115, 116
STERNUM: the entire ventral division of a body segment. Fig. 6.
Sthenarus, 199
stickseed, 116
Stictopleurus, 68
Stilt Bugs, 99
stink bugs, 40, 43, 46, 48, 50, 51, 53, 54, 55, 56
Stiretrus, 42
Stittocapsus, 168
strainer, 8
Strathmore board, 12
strawberries, 82, 87
STRIATE: marked with parallel fine lines.
STRIDULATORY: connected with stridulation; the production of sound usually by a specialized area or structure that is rubbed by an associated part. Fig. 2.
striola, Neoplea, 230
striolate, 159, 162
Strobilocapsus, 190
stygica, Nerthra, 244
stygicus, Slaterocoris, 175
STYLETS: mandibular and maxillary mouthparts enclosed by the labium. Figs. 4, 5.
Stylosanthes, 115
styrofoam, 15
suavis, Pallacocoris, 165
subcoleoptrata, Nabicula, 136
subglaber, Semium, 176
subis, Progne, 142
SUBMARGINAL: that part just within the margin.
SUBSTRATE: a part or area which lies beneath.
succinctus, Largus, 97
sugar cane, 209
sugar maple, 159
sulcata, Brochymena, 47
sulcifrons, Neottiglossa, 53
SULCUS: a furrow or groove.

Index and Glossary   255

sumac, 156
superbus, Adelphocoris, 170
SUTURE: a seam or impressed line.
suturellus, Dysdercus, 96
swallow bug, 140
swamp privet, 112
sweet clover, 170
sycamore, 79, 108
sylvestris, Ligyrocoris, 85
SYMPATRIC: two or more species inhabiting the same geographic area.
SYNONYMY: two or more names given for the same thing.
Synxenodorus, 142
Systelloderes, 117

T

Taedia, 164
takeyai, Stephanitis, 115, 116
Tannerocoris, 197
tape recorder, 8
Tarida, 139
tarnished plant bug, 1, 161
tarsalis, Corisella, 237
tarsalis, Trigonotylus, 153
TARSI: the segment attached to the distal end of the tibia. Fig. 1.
taurus, Repipta, 126
Taylorilygus, 163
tea strainer, 10
Teleonemia, 113
Teleorhinus, 193
temnosfethoides, Calliodis, 145
tent caterpillars, 1, 46
tenuicornis, Cylapus, 186
Teratocoris, 150, 152
TERETE: cylindrical or nearly so.
TERGITE: a dorsal sclerite or part of a segment.
tergum, 6
terminalis, Acanthocephala, 58
TERMINAL PLATE: a distinct convex area or lobe located on the distal end of the fore and sometimes the middle tibiae of some Reduviidae.
Tetraphleps, 146
Tetyra, 44
texanus, Renodaeus, 172
Thasus, 57
Thaumastocoridae, 25, 31, 98
thomsoni, Scolopostethus, 93
THORAX: the portion of the insect body bearing the wings and true legs.
Thread-legged Bugs, 119, 120
Thyanta, 51-52
Thyreocoridae, 22, 29, 37
TIBIA: the leg segment attached to the distal end of the femur. Fig. 1.

tick trefoil, 116
Tilia, 109, 111
tiliae, Gargaphia, 111
tiliae, Neolygus, 163
tiliae, Neurocolpus, 163
timothy, 53
tinctoria, Baptisia, 116
Tingidae, 9, 17, 25, 31, 106
tipuloides, Leptocorisa, 64
toad bug, 243
toe biters, 239
tomatoes, 102
Tominotus, 36
Torre-Bueno, Jose Rollin de la, 17
touch-me-not, 162
TRANSVERSE: at a right angle to the longitudinal axis of the body.
TRAPEZOIDAL: a four-sided configuration in which two sides are parallel and two are not.
Trapezonotus, 92
traumatic insemination, 139, 143
trays, 15
Trepobates, 214
Triatoma, 121, 130
trichobothrial spots, 4
Trichocorixa, 232-33
Trichopepla, 54
trifolia, Staphylea, 107
Trigonotylus, 152-53
trimaculata, Cryphula, 95
tripunctatus, Lygaeospilus, 73
trispinosum, Stenodema, 150
tristicolor, Orius, 146
tristigmus, Euschistus, 56
tristis, Anasa, 63
trivittatus, Leptocoris, 68
TROCHANTER: the leg segment between the coxa and femur. Fig. 2.
Tropidosteptes, 158
true bed bug, 140
true bug, 1, 2
true plant bugs, 205
TRUNCATE: cut off squarely at the tip.
trypanosome disease, 120
TUBERCULATE: covered with tubercles (small solid pimples or buttons).
Tuponia, 203
turbaria, Saldoida, 221
turcicus, Lygaeus, 71-72
turtle bugs, 40, 42
Tussilago, 165
TYLUS: the distal part of the clypeus between the juga. Figs. 1, 2, 4.
Tytthus, 203, 204

U

Uhler, Philip Reese, 18
uhleri, Barce, 132

uhleri, Cryptostemma, 207
uhleri, Lethocerus, 240
uhleri, Notonecta, 227, 228
uliginosus, Geocoris, 75
ulmi, Corythucha, 109
Ulmus, 109
ultra-violet lamp, 8
Umbelliferae, 161
umbelliferous, 54
umbrosus, Atrazonotus, 91
undata, Neottiglossa, 53
undulata, Notonecta, 227, 228, 229
unica, Diphleps, 205
unifasciata, Notonecta, 228-29
unique-headed bugs, 117
unus, Drymus, 95
ursinus, Pycnocoris, 157
Usinger, Robert Leslie, 18
usingeri, Cryptostemma, 207

V

vagans, Ceratocombus, 206
Van Duzee, Edward Payson, 18
vanduzeei, Rhinocapsus, 203
variegata, Derophthalma, 157
variegata, Physatocheila, 111
variolarius, Euschistus, 55
vegetables, 161
Veliidae, 21, 28, 214
velvet water bug, 218
velvety shore bugs, 244
VENTER: the undersurface of the abdomen.
VENTRAL: pertaining to the under surface.
ventralis, Rhynocoris, 122
venusta, Corythaica, 110
verbasci, Campylomma, 200
VERTEX: the top of the head between the eyes posterior to the front. Fig. 4.
verticalis, Trichocorixa, 232, 233
Vespertilionidae, 141
VESTITURE: covering; the general surface covering such as hairs, scales, etc.
vials, 8, 13
vicarius, Emblethis, 94
vicarius, Oeciacus, 140
vicinum, Stenodema, 150
vigilax, Neoborops, 160
vincta, Pachybrachius, 87
virescens, Cymus, 81
virescens, Orthotylus, 185
virginiana, Clematis, 179
virginicus, Chionanthus, 112
viridicans, Dichrooscytus, 169
viridicatus, Stictopleurus, 68
viridula, Nezara, 50
virilis, Corimelaena, 40
vitripennis, Hyaloides, 188
VITTA: a broad longitudinal stripe.
vittiger, Chelinidea, 59

vulgaris, Hesperocorixa, 235
vulnerata, Callicorixa, 236

W

walking sticks, 209
walnut, 109
wasp mimicry, 63
wasps, 118
water bees, 224
water bug, 242
water scorpion, 238
water spiders, 210
water striders, 210
water treaders, 217
weed flowers, 61
weedy fields, 61
western bat bug, 141
wheat, 56
wheel bug, 123
wherrymen, 210
white ash, 158
white birch, 41
white clover, 179
white-throated swift, 142
wickhami, Jalysus, 102
wild cherry, 108, 109
wild cucumber, 63
wild four o'clock, 62
wild geranium, 167
wild gooseberry, 167
wild indigo, 116
wild onion, 179
wild parsnip, 162
wild raspberry, 203
willows, 108, 110, 111, 146, 160, 173, 196, 202
woodpecker, 142
woodpecker holes, 142

X

Xenoborus, 158
Xenogenus, 67
Xestocoris, 90
Xylastodoris, 98
Xylocoris, 145
Xyonysius, 79-80

Y

yellow columbine, 101
yucca, 191

Z

Zelus, 124
Zeridoneus, 86

NOTES

NOTES

NOTES

COLLEGE OF MARIN LIBRARY

3 2555 00030680 8

QL
522.1
A1
S53

Slater, James Alexander

How to know the true bugs (Hemiptera-... 57996

DATE DUE

MAY 1 4 1992			
MAR 1 0 1994			
OCT 2 0 1984			
OCT 1 7 1996			
DEC 1 3			
APR 2 4 2000			
MAY 2 5 2000			

COLLEGE OF MARIN LIBRARY
KENTFIELD, CALIFORNIA

COM